T0342156

Handbook for Process Plant Project Engineers

by

Peter Watermeyer

Professional Engineering Publishing Limited
London and Bury St Edmunds, UK

First published 2002

ISBN 1 86058 370 9

A CIP catalogue record for this book is available from the British Library.

Contents

Related Titles

How Did That Happen – Engineering Safety and Reliability	W Wong	1 86058 359 8
IMechE Engineers' Data Book (Second Edition)	C Matthews	1 86058 248 6
Improving Maintainability and Reliability through Design	G Thompson	1 86058 135 8
A Practical Guide to Engineering Failure Investigation	C Matthews	1 86058 086 6
Process Machinery – Safety and Reliability	W Wong (Ed.)	1 86058 046 7
Plant Monitoring and Maintenance Routines (IMechE Seminar)		1 86058 087 4

For the full range of titles published by
Professional Engineering Publishing contact:

Marketing Department
Professional Engineering Publishing Limited
Northgate Avenue
Bury St Edmunds
Suffolk
IP32 6BW
UK
Tel: +44 (0)1284 724384
Fax: +44 (0)1284 718692
Website: www.pepublishing.com

Foreword

Project engineering comprises the orchestrated integration of all engineering disciplines to achieve a commercially defined project objective. The quality and effectiveness of project engineering are key components of process plant design and construction, a business which invariably requires complex multi-disciplinary design input and focused customer and commercial awareness.

Courses in the individual disciplines such as civil, mechanical, and chemical engineering are offered by many universities and other institutions. By contrast, the art of project engineering is usually acquired the hard way, through practical involvement in project execution. Unfortunately, much of the necessary experience is too often painfully gained by learning from expensive mistakes.

Peter Watermeyer has a keen interest in analysing the causes of success, failure, mediocrity, and excellence in project execution. He sees these as a function of not only the performance and management of the engineering work, but also of the way that work and project teams are structured and orientated in relationship to their whole environment. Peter has a wealth of practical experience in the industry. He spent 15 years in the power, hydrocarbons, and process machinery industries before joining Bateman, where he has spent over 20 years in the engineering and management of process plant projects, mainly in the international mining and petroleum industries.

This book, which is based on the knowledge, skill, and diverse experience of a master of the art will, I believe, prove to be of great value in guiding aspiring and less experienced project engineers towards achieving full professional competence in this challenging arena.

Dr John P Herselman, DrIng, Dipl-Ing, BSc (Chem Eng), FIOD
Executive Chairman, Bateman BV

Acknowledgements

The illustrations in this book are by Natalie Watermeyer and the photographs and drawings are reproduced with permission from Bateman Projects Ltd.

I am grateful to Phillip Ashworth, for his review and guidance on the revision to the original text. I would also like to thank the editorial staff of Professional Engineering Publishing, and in particular the commissioning editor Sheril Leich and the co-ordinating editor Mandy Kingsmill, for their very professional work.

Chapter 1

Introduction

This book is intended to assist people who design and build process plant, and people who participate in multi-disciplinary engineering projects in general. The book is aimed in particular at the project engineer, or team of lead engineers, who inevitably hold such projects together, being at the heart of the information generation system which shapes and guides the project.

The design and construction of process plant covers an exceedingly wide field of performance and knowledge. Considering engineering topics alone, those which impact process plant design include the work of very many specialist technical branches. But this is only part of what has to be considered in successful project engineering. Interwoven with the engineering and design of the plant are many commercial, constructional, financial, and social considerations. It may be satisfying for the technical purist to focus exclusively on engineering issues, but the engineer who does so is likely either to be limited to a subordinate role, or to be part of an unsuccessful enterprise – engineering considered on its own becomes an academic pastime.

The content of the book is therefore intended to address that mixed bag of technical, commercial, managerial, and behavioural issues which constitute the actual job content of the practising engineer, and are peculiar to this industry.

1.1 What's so special about process plant projects?

In some cases – nothing! A relatively small and simple project can usually be executed by a single experienced engineer with a minimum of fuss, and without complex procedures or strategy. This is especially the case for plants which are very similar to existing units. Within the following text may be found some helpful aspects of technical advice for such projects, but the work must not be over-complicated. Any higher-level issues of strategy and management are best addressed (if at all) at the level of organization which encompasses the small project, for example the corporation which requires the project, or a contracting organization which delivers such projects. The following text is primarily addressed at relatively large and complex projects, requiring the interaction of a number of skills – technical, commercial, and constructional.

Compared to the general field of engineering design and construction, the main differentiating factors which have to be addressed in these projects are the following.

- The unique design of each plant is the inevitable consequence of the need to optimize each application to its unique circumstances of feedstock, product, capacity, and environment.
- Plants are built around hundreds of items of proprietary processing equipment. The plant design must interface exactly with the operational characteristics and dimensions of these individual items. The interface information can only be finalized when the commercial agreements with equipment suppliers have been concluded.[1]
- Both the plant design and its construction employ many types of specialist, who must interact at thousands of interfaces (and often work simultaneously in the same plant space envelope) to produce a co-ordinated product.
- Plant operation can be hazardous. The elimination or control of hazards, and the establishment of safe operating practices, are prerequisites of plant operation, requiring priority attention during plant design. Environmental impact is also invariably an issue.
- Technology development is rapid and continuous. This greatly impacts on the uniqueness of process and plant design (the first item above), but also often impacts on project execution – there is often a desire to accommodate changes while design and construction are in progress.

---

[1] This is at least the common practice in free-market economies. In communist countries, it was (and still is) often the practice to 'centralize' equipment designs, so that the plant designer can simply choose a completely detail-designed equipment item from a catalogue. This simplifies the plant designer's job,but often at great cost at the construction stage, when there is no commitment from a supplier to match the project programme or budget. Also, it shifts responsibility for 'fitness for purpose' of the item.

- The project schedule cannot be generated in any mechanistic fashion. Invariably, the critical path can be shortened almost *ad infinitum* by taking various shortcuts (in other words, by changing the schedule logic – not just by employing more resources or working faster). These possible shortcuts come at a cost or risk, which must be balanced against the benefits.
- The nature of the industry served by process plant usually puts a high premium on early plant completion and operation.
- The plant has to be constructed on site, to suit its site, wherever that may be.

The degree to which these and other features are present, together with the more obvious comparisons of size, cost, and complexity, dictate the intensity of the challenge to the project team. All of these features have to be addressed during the plant engineering and its follow-up.

1.2 The structure and components of this book

The main objective of this book is to give guidance on how the project engineer's work is carried out in practice. We have therefore to consider how all items of information interact, and how they are brought together in each practical situation. This is a distinctly different objective from that of a technical reference book, such as Perry's *Chemical Engineer's Handbook*. This is a recommended reference for any engineer in the process plant field, and contains a wealth of information, but it is targeted and indexed by technical subject. We need a different structure here, in which the sequence of information is important and is project-related. It is convenient to address the structure at four levels.

In defining these levels, an immediate pitfall can be identified and should be avoided. In organizational and management terminology, invariably there is a high and a low level. This should not be confused with relative importance – all of the organization components are important. *Within* each level, there are likely to be relatively less and more important functions for a given project, but a project can be jeopardized just as easily by poor detail design as by poor management.

Subject to this qualification, the base tier, level one, consists of the activities of detailed design, procurement of parts, delivery to site, construction, and commissioning, and all of these will be addressed. Inevitably in an integrating text such as this, most topics have to be discussed at a relatively superficial level, concentrating on those facets which are more essential, either to the interactions with the whole, or because of relatively high individual impact to the project and the plant.

At the level above, level two, reside the technical focus and project control, which give direction to all the detailed work. The centrepiece of this is the process technology package. The technological content of such packages is outside the scope of this book – it is the domain of many experts and organizations, embracing diverse fields of technical specialization such as catalytic conversion and hydro-metallurgy, and niches within these fields which yet consume entire careers. We will rather (and briefly) address the generalized make-up of such packages, of how they relate to the plant to be built and to the project around them. In addition, we will discuss the system of engineering and information management (also at level two), which governs the performance of level one activities.

Not particularly addressed at level two, because of the focus of this book, are the methodologies of managing procurement, logistics, contracts, finance, and construction. The reader with greater project management aspirations is advised to refer to specialized texts on these subjects. We will, however, discuss the principal interfaces with engineering work and its management.

At level three, we have a management system for the entire project. This is conventionally broken down into three or four components, namely management of scope, quality, cost, and schedule. (Scope and quality may be regarded as a single issue, a practice not recommended by the author.) Health and safety considerations may properly also be managed at this overview level, and must be included in any such text.

At level four, there is only one item, project strategy. This has to ensure that the project is correctly conceived (technically, commercially, economically, socially) and embodies the skeleton of the over-arching plan which will ensure that the goods are delivered in the optimum fashion. Inevitably the strategy must deal with the issues of relationship management between the principal stakeholders, for example the plant owner (usually, a complexity of people and interests) and contractors, including possibly a single lump sum or managing contractor, and a variety of sub-contractors and suppliers. The strategies are bound to be different for each stakeholder, reflecting the basic question of 'What do I want out of this project?'

There is a significant body of professional opinion that such considerations of strategy are not appropriate in the context of what should be the technical field of engineering. (In the author's experience, there are also several senior executives who are relatively ignorant of the links between engineering and strategy, and feel quite threatened!) There are those who argue that management of large projects, whether or not centred on multi-disciplinary engineering, is primarily a matter of project management practice, and can be quite separately treated as such. Both of these views can be dangerous. From level one to level four, there are common threads, mainly of engineering information and its development, which have to be taken into account when making management decisions. The understanding and maintenance of these threads, many of whose origins are at the detail level, are the essence of the discipline of project engineering.

The interaction between project strategy and engineering execution operates in two directions. It will be seen within this text that the adoption of project strategy – including for instance the structure of the relationship between client and contractor – has major implications for the way in which engineering is optimally conducted. In reverse, the best project strategy (for each party) is likely to depend on some of the engineering realities.

The technical content of this book posed a particular challenge of selection and condensation. Nearly every topic could be expanded into several volumes, and must in practice be covered by specialist engineers with far more knowledge than is presented. The intentions behind the information presented are twofold. Firstly, to include just sufficient information for a generalist (the project engineer) to manage the specialists, or to assist the specialists to co-ordinate with each other and find the best compromises. Secondly, to include some of the author's own experience (spanning some 30 years) of simple but important items which do not seem to be conveniently presented in textbooks or standards, but which can prove important and costly.

1.3 Methodology of presentation

Project engineers have traditionally learnt much of their job-knowledge by experience. The mainly technical aspects can be taught in isolation as an academic subject, but even these have to be tempered by experience, to match theory to application. The behavioural and management aspects of job performance are even more dependent on the acquisition of relevant experience. Through a few project cycles in relatively junior roles, engineers have gradually broadened their skills and their understanding of what is going on, inside and outside of the project team, and become aware of the consequences of design decisions and of the way actions and events interact. They have learned to recognize problems, and learned some standard solutions and how to choose the best solution for the circumstances. They have also absorbed something of the culture of their industry, and learnt how to behave and influence others, to make the best of their working relationships. They have become ready for a leading or management role.

It is our objective here to produce a text which accelerates this process of learning, and enhances it by trying to transfer some of the author's own experience and retrospective analysis of successes and failures. Surely the process of learning by experience can be accelerated, and some of the pain reduced.

A limitation of trying to synthesize such a learning process is that a balanced progression must be achieved, in which all the interacting subjects are advanced more or less in parallel. It is no good giving a dissertation on the merits and techniques of compressing project schedules unless the more normal schedule practice and its logic are understood. It is equally fruitless (and boring, and unlikely to be absorbed) to discuss the finer points of project management, let alone strategy development, until it is understood what are the work components (and their characteristics) which have to be managed and optimized.

The book has therefore been set out into six cycles, each targeted at a different and progressive stage of development. The first cycle consists of an overview of a process plant, the process technology package, a very brief outline of the management of a project to build a plant, and a brief description of the engineering work and its management. The second cycle looks at the project environment, in particular some non-engineering factors which influence the work of the project. The third cycle is about project initiation and conceptual development. This becomes possible when the nature of the project and its environment are understood. We have also chosen to address the subject of hazards and safety at this point. The fourth cycle addresses plant engineering technical issues, at a more detailed level, and is the largest component of the book. The fifth cycle is concerned with project engineering and management issues which need a particular emphasis in this industry, leading up to a final cycle which discusses strategy development.

It may be noticed that the two structures, of level and of cyclical development, are not strictly parallel, but compare more like a traditional four-cylinder engine firing cycle of 2-1-3-4. Even this analogy is somewhat misleading. There is no single direction leading to strategic understanding, but rather a matrix where every part depends more or less on every other part, and the strategy requires a knowledge of all the parts. Strategic performance, without detailed performance, is of no value whatever, and both continue to evolve and require a continuous learning process.

Turning to the basics of the format of presentation, in an endeavour to provide a text that can be useful for on-the-job reference, the general mechanism employed is the checklist. Wherever possible, the text has been formatted into bullets, which are intended to provide a structured approach to the common practical situations, by helping the engineer both to plan his[2] approach and to check that he is addressing all essential facets.

1.4 Getting it right

The basic project methodology advocated in this book is to start a project by looking as widely as possible at the choices presented, evaluate the full consequences of those choices, and then develop a strategy to make the best of the opportunities and eliminate the risks. The strategy is further developed in a structured fashion, leading to detailed plans and methodologies, which are based on solid experience that has been proven to give manageably acceptable results.

It all sounds very simple. And yet, so many individual engineers, and organizations of engineers, fail to produce acceptable results. Individual intelligent and well-educated engineers fail to rise above mediocrity, and whole engineering organizations become obsolete and disappear. Others prosper: some seemingly by sheer luck, by being in the right place at the right time, but mostly by getting the right balance between having a good strategy, based on recognition of personal objectives, and the diligent deployment of sound skills.

There really is not much more to be said about the path to success, but it is quite easy to comment on the failures. A few main categories are worth mentioning.

- *Bad luck.* As real calamities such as political risk or natural disaster can and should be insured against, this should rather be described as the occasional, unfortunate consequence of excessive risk. Some degree of technical risk is inevitable for developing technologies, without which a technically based enterprise is anyway unlikely to flourish. The same goes for commercial risk, which is necessary to seize opportunities. The key is to have a considered risk management

[2] Throughout the text, the male gender is intended to include the female. This is done purely for the purposes of simplification.

Fig. 1.1 '... a good strategy, based on the recognition of personal objectives, and the diligent deployment of sound skills'

plan, which recognizes and excludes risks whose possible consequences are unacceptable, and balances manageable risk against reward. This should lead to ample compensation in the long run. In conclusion, this is an element of strategy.

- *Lack of knowledge* (job-information, know-how or commercial practice) or of diligence. The following text should hopefully impart enough knowledge at least to assess how much additional expertise and information is required for a given project. No remedy is available for the second shortcoming.
- *Setting of unreasonable targets.* The detail of how this sometimes comes about, and how it can be recognized, will be addressed in the text – it is really a relationship issue. Occasionally, social or political factors, or 'brain-dead' enterprise directors, spawn projects which inherently have little or no real economic return – a conclusion which is hidden by an unreasonable target. Stay clear of these projects!
- *Running in reactive mode.* Here is an example. A natural resources company needs a new process unit to enhance its product slate. Its technical manager (the client) is appointed to oversee the project. He is a leading expert on the technology application, and Knows What He Wants, but is not terribly good at communicating it – or listening for that matter. In due course a project team is assembled under a project manager, including a process design team and a lead engineer

for each discipline. A planner is recruited. He loses no time in getting plans from each discipline, and putting them together into an overall network. Unfortunately, he does not fully understand the interaction between disciplines of detailed engineering information, the potential conflicts, or how they may be resolved. The client dictates that the project schedule must be 16 months to commissioning, because he heard that was achieved somewhere else, and the planner duly jockeys the project schedule around to reflect this. The client also dictates the project budget. This includes no contingency, because he Knows What He Wants, and contingencies will only encourage the project team to believe that error and waste are acceptable.
As time goes on:

- Engineering gets further and further behind schedule, as each discipline in turn is unable to start or complete jobs because of the unavailability of information from equipment vendors and other disciplines.
- Design and procurement decisions are endlessly delayed, because the client is not offered what he Wants, and expects the project team simply to make up the delays. They do not.
- Many of the project drawings become full of revisions, holds, and – inevitably – interface errors, because the wrong revisions of input information were used, or assumptions were made but not verified and corrected.
- In the struggle to catch up, quality functions, such as layout design reviews and equipment and fabrication inspections, are skimped.
- The project manager loses control of the project – in fact, he spends most of his time arguing with the client, usually to try to stop process and layout changes being made. ('These aren't changes', the client would say. 'They are corrections of error. You failed to design what I Wanted.')
- The construction site degenerates into chaos. The construction management are unable to handle all the late drawings, design errors, late deliveries, flawed materials, and acceleration demands.
- Both schedule and budget are grossly overrun, and the construction contractors make a fortune out of claims, especially for extended site establishment. And the plant is *not* What The Client Wanted.

If none of this seems familiar – you are new to process plant work. But it is all quite unnecessary.

First Cycle

A Process Plant and a Project

Chapter 2

A Process Plant

2.1 Basic process design elements

A process plant is a classification of factory which transforms materials in bulk. The feedstock and products may be transported by pipeline or conveyor, or in discrete quantities such as truckloads or bags, but they are recognized by their bulk properties. Examples of process plants are oil refineries, sugar mills, metallurgical extraction plants, coal washing plants, and fertilizer factories. The products are commodities rather than articles.

The plant consists of a number of the following.

- *'Process equipment' items*, in which material is transformed physically or chemically, for example crushers, reactors, screens, heaters, and heat exchangers. The process equipment is required to effect the physical and chemical changes and separations necessary to produce the desired products, and also to deal with any unwanted by-products, including waste, spillage, dust, and smoke.
- *Materials transport and handling devices*, by which the processed materials and effluents are transferred between the process equipment items, and in and out of the plant and any intermediate storage, and by which solid products and wastes are handled.
- *Materials storage facilities*, which may be required to provide balancing capacity for feedstock, products, or between process stages.
- *'Process utilities'* (or simply 'utilities'), which are systems to provide and reticulate fluids such as compressed air, steam, water, and nitrogen, which may be required at various parts of the plant for purposes such as powering pneumatic actuators, heating,

cooling, and providing inert blanketing. Systems to provide process reagents and catalysts may be included as utilities, or as part of the process.

(Note: All of the above four categories include items of mechanical equipment, namely machinery, tanks, pumps, conveyors, etc.)

- *Electric power reticulation*, for driving process machinery, for performing process functions such as electrolysis, for lighting, for powering of instrumentation and controls, and as a general utility.
- *Instrumentation*, to provide information on the state of the process and the plant, and, usually closely integrated to the instrumentation, control systems.
- *Structures* (made of various materials, including steel and concrete), which support the plant and equipment in the required configuration, enclose the plant if needed, and provide access for operation and maintenance.
- *Foundations*, which support the structures and some plant items directly, and various civil works for plant access, enclosure, product storage, and drainage.
- *Plant buildings* such as control rooms, substations, laboratories, operation and maintenance facilities, and administration offices.

In addition there are inevitably 'offsite' facilities such as access roads, bulk power and water supplies, security installations, offices not directly associated with the plant, and employee housing; these are not considered to be part of the process plant.

A process plant is fundamentally represented by a process flowsheet. This sets out all the process stages (essentially discrete pieces of process equipment) and material storage points, and the material flows between them, and gives corresponding information on the flowrates and material conditions (chemical and physical). This information is usually provided for:

- the mass balance case, in which the mass flows will balance algebraically;
- a maximum case, corresponding to individual equipment or material transport maxima for design purposes (these flows are unlikely to balance); and
- sometimes, by cases for other plant operating conditions.

For thermal processes, the mass balance may be supplemented by a heat and/or energy balance.

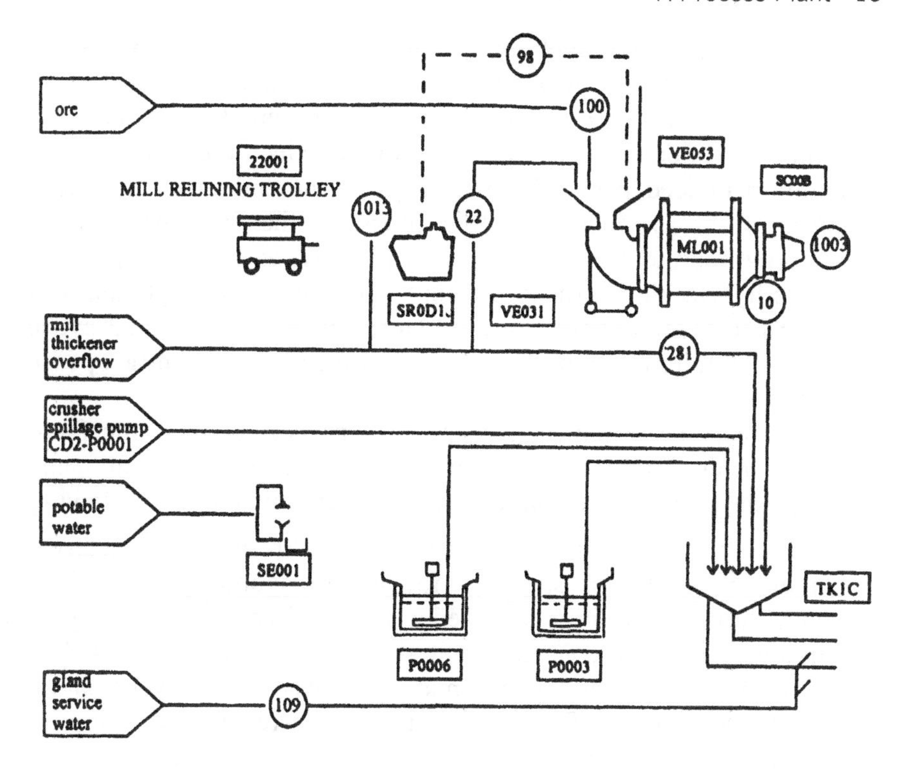

STREAM NUMBER	10	22	98	100	109	281	103	1013
DESCRIPTION	MILL O/F SLURRY	MILL DILUTION SOLUTION	MILL STEEL BALL CONSUMPTION	ORE FEED TO MILL	G.S.W. TO MILL DISCHARGE PUMP	MILL SUMP DILUTON SOLUTION	TROMMEL SCREEN O/SIZE	SCREEN SPRAY WATER
TOTAL FLOW t/h	700	450	0.410	200	15	500	-	900
TOTAL FLOW m3/h	500	400	0.100	100	15	480	-	900
SOLIDS FLOW t/h	240	0	0.410	200	0	0	-	0
SOLIDS DENSITY t/m3	2.90	-	4.10	2	-	-	-	-
OTHER PHYSICAL PROPERTIES	↓	↓	↓	↓	↓	↓	↓	↓

Fig. 2.1 Flowsheet with mass balance

The process flowsheets represent the process rather than the details of the plant. The latter are shown in 'P&I'[1] diagrams, which depict all

[1] Pipeline and instrumentation, although sometimes described as process and instrumentation; but P&I has become an accepted international multilingual expression. Some engineers use 'mechanical flow diagrams', which do not show much instrumentation, and 'control and instrumentation diagrams', which focus as the name implies, and no doubt such presentation is appropriate for certain applications; but P&IDs usually suffice.

plant equipment items, including their drive motors, all pipelines and valves (including their sizes), and all instruments and control loops.

Utility flowsheets and diagrams are often presented separately.

Plants may operate by batch production, in which the plant processes a quantum of feed per cycle, and stops at the end of each cycle for removal of the product and replacement of the feed. Alternatively, plants may operate continuously, 24 h per day, without stopping; and there are hybrid plants, or hybrid unit operations within plants, which are described as semi-continuous, in that the internal operation is cyclical but the cycles follow continuously, one after the other, with little operator intervention.

The critical performance factors for a process plant – the factors which determine its fitness for purpose and its effectiveness (and against which its designers' performance is measured) – include the following.

- *Feedstock transformation as specified.* Product characteristics should be within a specified range corresponding to feedstock characteristics within a specified range, and capacity (throughput) should be within the range required for feed and for product. From the feed and product capacity may be derived the recovery, or yield of product per unit feed. Alternatively, the recovery and input or output may be stated, and the output or input respectively may be derived.[2]
- *Cost of production*, often expressed per unit of feed or product. The cost components include capital amortization and interest, plant operators' salaries, maintenance materials and labour, purchased utilities, process reagents, insurance, etc. There may also be fees payable to process technology licensors. The capital cost component is often quoted separately as a stand-alone criterion.
- *Plant reliability and availability.* Reliability is the predictability of plant operation as planned, whereas availability is the proportion of time for which the plant is in a condition whereby operation (to acceptable standards) is possible. Availability may be less than 100 per cent because of planned outages for maintenance, for example one 3-week shutdown per 2-year cycle, or because of shutdowns caused by lack of reliability, or (invariably to some degree) both.
- *Safety of construction and operation.* This is assessed at the design stage by formalized hazard analysis for the process and by hazard

[2] For some plants, there may also be a specification linking the feedstock or product specification (or grade) to the capacity and/or the recovery.

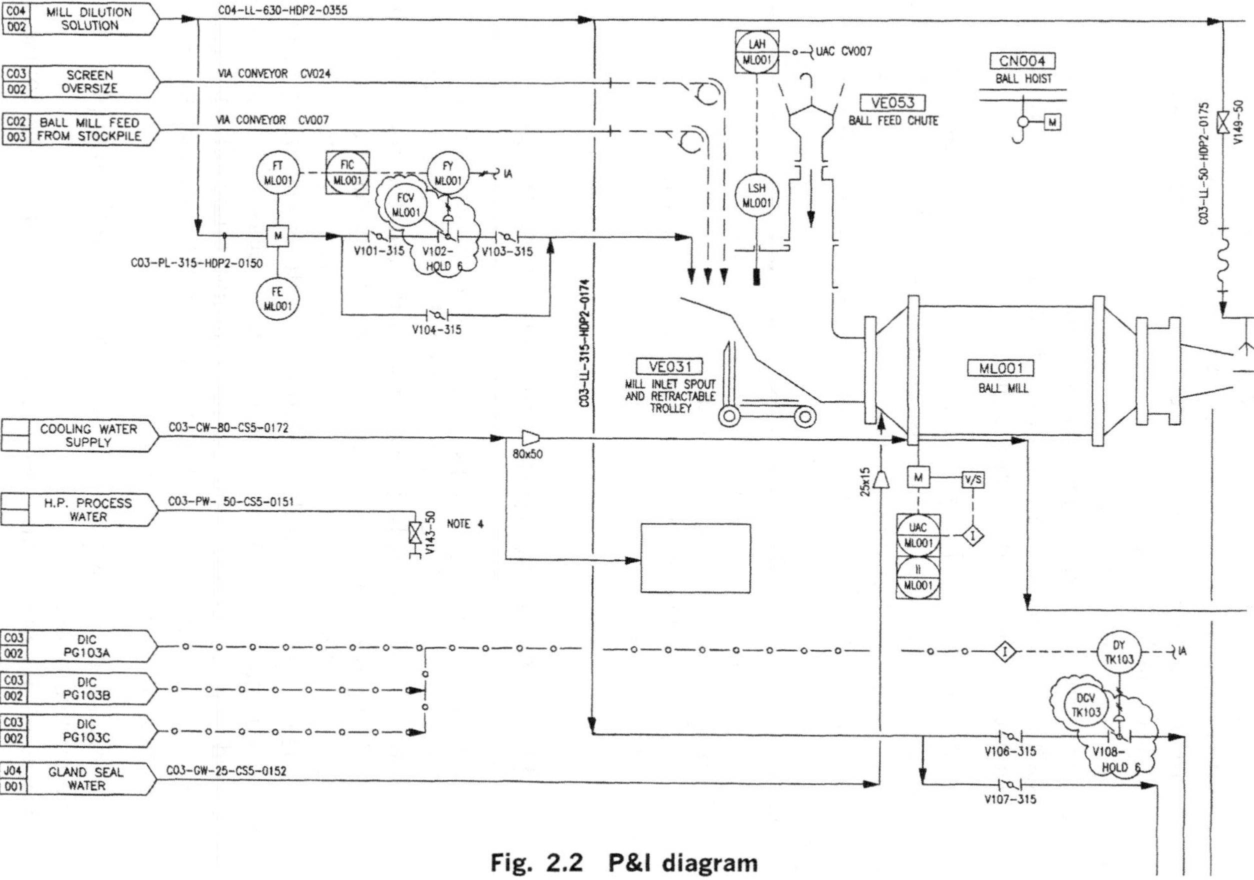

Fig. 2.2 P&I diagram

and operability (Hazop) study of the plant design. It is assessed during construction and operation by audit of the presence and efficacy of various safety features and constructional and operational practices, and it is reported historically by accident and loss statistics.

- *Environmental impact*, and its acceptability by legal, social, and ecological consideration.
- *The plant life.* Plant maintenance practices and costs are presumed for the purposes of economic analysis, and hopefully in practice, to be such as to keep the plant operational within specified performance levels over the intended life of the plant.

The last four factors clearly have an effect on, and their economic impact should be included in, the cost of production. However, they are important design and evaluation factors on their own, and may also have an effect on product marketability, or in some cases whether the plant is permitted to operate at all.

The time taken to build and commission the plant, and get it into full commercial operation, is equally a factor which significantly impacts on the planned and actual cost of production over the plant life. It may also have an important effect on the marketing, and hence economic value, of the product.

2.2 The processed materials and the process

For the purpose of classifying types of plant in order to observe some of the principal features of each type, the first differentiating features are the nature of the materials to be processed, and the usage of the product. The processed materials may be principally classified as fluids or solids, hazardous or non-hazardous, and minerals, bio-matter or water. The main types of product addressed are fuels, chemicals, metals, precious minerals, and foods. Usually, each feedstock is primarily associated with a certain product, for example crude oil with fuels, and bio-matter with foods and pulp products. Of course, there are combinations of these groups, such as:

- processed materials which begin in a predominantly solid phase and end in fluid phase;
- processed materials which become hazardous during the course of processing, such as explosives;
- foods (such as table salt) which are minerals;
- fuels extracted from bio-matter; and

- materials, such as methanol, which may be of mineral or vegetable origin, and hopefully not regarded as a food.

Many materials and products, such as pharmaceuticals, will not be addressed specifically in this book, being too specialized in nature, and the reader interested in such processes will have to draw his own conclusions as to the relevance of these contents.

Processes may be classified according to:

- the complexity (usually quantified by the number of items of process equipment);
- the severity of the associated physical conditions (including pressure, temperature, and corrosiveness – clearly, in conjunction with the processed material characteristics and toxicity, these impact on the degree of hazard);
- whether there is a continuous or batch process; or
- the state of evolution of the technology employed.

Some processes have very critical product specifications, especially as regards the permissibility of contaminants, and an understanding of how to manage the production process accordingly is essential to success in designing and building such a plant. A plant which is designed to produce salt as a food requires different considerations from a plant which produces salt as an industrial chemical.

There is a particular purpose to the foregoing, and that is to alert the reader that both the feedstock and the utilization of the product are critical to choice of process and plant design. Industries tend to develop a processing methodology and design practice which is appropriate for their normal feedstock and product utilization. If either is changed, both the process and plant design practice must be critically examined, although the client may be unaware of the need (taking it for granted that the requirements are known).

Some processes are simple enough in concept and easily managed in practice, and the criteria for building the corresponding plant are too obvious to need much elaboration. However, in general, the plant design needs to embody much more information than is contained in the flowsheets and the P&I diagrams. The actual additional information required is peculiar to each process, and it should be understood that no list of such information headings can be comprehensive. For instance, it is sometimes found that a relatively small equipment detail, such as a sealing device, or a feature that avoids the accumulation of material build-up, is critical to the operation of a whole plant, and is central to the initial development of the process as a commercial enterprise.

The following are some general process information requirements, in addition to plant performance specifications, flowsheets, and P&I diagrams, for designing a plant.

- *Data sheets for each item of equipment* (including instruments), detailing the performance requirements, the environment (including the range of process materials and physical conditions encountered, with emphasis on harmful conditions such as corrosion, abrasion, or vibration), the materials of construction, and any special design features required.
- *General specifications for the plant and its components*, embodying the particular requirements for the process (such as corrosion resistance, hazard containment, or features to promote reliability), and including material and valve specifications for piping systems, and specific instrumentation details.
- *A description of the method of plant operation*, including start-up, shutdown, management of predictable plant operational problems, and emergency shutdown if applicable.
- *A description of the hazards inherent in the process and plant operation*, and the safety features and precautions to be taken to overcome them. This may include the classification of hazardous areas.
- A narrative supplementing the P&I diagrams to describe the *method of controlling the process* and the plant, and the instrumentation and control system architecture, usually known as the 'control philosophy'.
- In the case of products such as foods, *reference to the regulations and requirements* of the appropriate food and drug administrations, and the detailed processing features required to meet them.
- Usually, *a basic layout of the plant.* This becomes essential if some of the layout features, for instance minimization of certain materials transport routes or maintenance of minimum clearances around certain items of equipment, are critical to plant operation or safety.

2.3 The process design/detail design interface

Quite commonly, the information described above is collectively referred to as the 'process package'. It may indeed be presented as such (for a suitable fee!), appropriately bound and decorated. We will discuss later the various facets of work organization which determine how work is packaged and how the process technology input can relate to the balance of engineering work, but for the moment, it should be appreciated that the engineering of the plant can be separated into two parts. The first part is the process technology which would be applicable to a plant built on any site,[3] and in general utilizing any combination of appropriate equipment vendors. The second part is the complete engineering design of the plant, incorporating actual proprietary equipment designs, locally available construction materials, local design practices and regulations, customized design features required by the particular client and his operation and maintenance staff, and layout and other design features necessary for the plant site. The second part is often referred to as the 'detailed engineering'. There is an area of potential overlap between the two; in particular, the process package can be expanded to a 'basic engineering' package, which includes the essence of detailed engineering (such as well-developed plant layouts and equipment lists) as well as the process package.

It needs to be understood that it is difficult to include all the required knowledge of a particular type of process into a stand-alone package. It is even more difficult in the case of complex processes or newly developed processes. In practice, when the process technology provider (or licensor) is separate from the detailed engineering organization, it is necessary to have most of the important design details reviewed and approved by the technology provider, to ensure that the process requirements have been correctly interpreted. Further process technology input is also needed in the preparation of detailed plant operation manuals and plant commissioning, and sometimes into the plant construction.

[3] This is an over-simplification. In practice, the process design package usually has to be customized to take into account local factors which affect the process, such as ambient conditions, feedstock and reagent variations, and properties of available utilities.

Chapter 3

A Project and its Management: A Brief Overview

The purpose of this chapter is to look briefly at all the components of a process plant construction project and its management, before focusing on the engineering work.

3.1 The project

We shall define a process plant project as: the execution of a plan to build or modify a process plant, within stated parameters of workscope, plant performance, cost, and time. This is not a universally accepted definition, but it does focus on the work with which the project engineer is principally engaged. Pre-project work will be described as a study or proposal.

Some project practitioners refer to the concept of a 'project lifecycle', which includes the initiation of a project, its technical and economic evaluation, funding and authorization, design and construction, plant operation, maintenance and further development, and finally plant decommissioning. We would rather call this a 'plant lifecycle', but the difference is mainly semantic: the various stages all have to be taken into consideration when designing and building the plant, which is our main concern and regarded here as being 'the project'.

Following on from the lifecycle concept, there is no clear-cut requirement on where we should begin our description of project work. The design of the plant depends *inter alia* on how it will be operated, the operation on how it was designed. The cost of the plant depends on how it is designed, the design depends on how much money is available to build it. The feasibility study has to anticipate the project outcome, the project is initiated with parameters set by the study. Because our focus is on the design and construction of the plant, we shall give minimal attention here to the work done prior to the decision to go ahead. These aspects are given more detailed review in Chapter 9, Studies and Proposals.

A project may be a major new enterprise such as an oil refinery, or a small modification to an existing plant. For the latter case, it is evident that the desired objective can be achieved quickly and informally without elaborate procedures. As the size and complexity increase, the need for a more formal approach becomes apparent, for many reasons. More people are involved, more interacting components have to be co-ordinated, and the investors demand more detailed reporting. The size of project at which more formalized procedures become necessary depends very much on the ability and skills of the project manager and the demands of the client. In general we will be addressing the needs of the large project, on the basis that engineers who are experienced in the bigger picture will understand what shortcuts and simplifications are reasonable on smaller projects.

Pre-project work starts with an idea or concept which the client has decided to develop. The concept and design of the final process plant progresses in cycles of increasing definition. Initially a study is made, in which the concept is technically developed, optimized, and analysed as a business proposition; the analysis includes considerations of technical and commercial risk, capital and operational cost, product value, and return on investment. A report is prepared; if the conclusions are acceptable to the client, he may authorize the implementation of the project. Alternatively, he may authorize more funds for further conceptual development, or, of course, abandon the concept. Authorization of the implementation of the project invariably implies the expectation of a plant which will perform within specified limits, and be built in accordance with certain standards, within a promised budget and schedule.

There is clearly an amount of pre-project engineering work necessary to achieve the required degree of technical definition, costing, and schedule analysis required for authorization. Prior to the decision to implement the project, there is a natural reluctance to spend any more funds than are absolutely necessary to complete the feasibility study, as there may be no project. This reluctance is tempered by the need for accuracy – evidently, the further the engineering of the plant is developed, the greater the confidence in the accuracy of the study report. In addition, if the client is confident that the study will lead to a project without much further conceptual design development, he may be willing to commit more funds prior to final authorization, in order to expedite the project.

In the review on engineering work which follows in the next chapter, for most activities we will approach the project work as though there were no preceding study, so that nothing is left out. In practice, a project invariably commences with a process of assembly and critical scrutiny of the preceding documentation. If the project team doubts the viability of the commitments made in the study, there may be some 'debate and compromise'. (This may be an understatement!)

3.2 Starting the project

We have established that a project commences with defined goals and a defined and accepted concept for achieving them. A project is not a process of innovation; the innovative work necessarily precedes the project, enabling the designs therefrom to be evaluated and costed. It may be found at the project commencement that the conceptual design is incomplete or unsatisfactory, that innovation is needed, or that plans or estimates are questionable. This is usually the case, to some degree. It is important to complete or rectify this work in an initial baseline-development phase, which on completion will permit the establishment of a detailed master plan of all project activities and cost elements, to be used as the basis of schedule and cost management. Without a fully developed baseline built on approved conceptual designs, the project manager can do no better to manage the project than a team can play competitive football without designated field markings and goals.[1]

Opportunities for improvement by innovation or by changing concepts may emerge during the project work, and if changes are made as a result, they have to be very carefully controlled if chaos is to be averted, though it is seldom acceptable to ignore such opportunities on principle. But these are exceptions: the objective for project work is to achieve the targets set out in advance. In proportion to the scale and complexity of the project, and therefore the number of different entities whose work has to be co-ordinated, is the need for detailed planning and control of the work done. This is succinctly expressed by American engineers as *'plan the work then work the plan'*.

'Plan' is used here in the widest sense, embracing not just the development of activity sequences and durations, but also the full evaluation of

[1] To be more accurate, it is not possible for the type of project work envisaged here, where a defined amount of work is required to be performed for a pre-determined cost. It is of course possible to carry out elastic-scope and undefined cost or schedule work without much of a baseline.

Fig. 3.1 'Plan the work then work the plan'

objectives and circumstances, potential problems, options, and methods available before undertaking any task. The plan must be verified to meet the project requirements before commencing the planned activities.

So we start the project with the preparation of a master plan that is verified to meet the project requirements, and is preferably formally approved by the client. If the client's exact requirements are already set out in a contract, as for a turnkey plant, the plan should be checked against the contract specification.

The plan includes the following.

- A clear definition of the project's end-product and of how its acceptability is decided, including:
  - the physical plant and its performance;
  - construction specifications and facilities;
  - documents required to operate and maintain the plant;
  - close-out reports and financial accounts;
  - services such as commissioning and training; and
  - a statement of any other client requirements to be met in the course of achieving the end-product, such as the format and content of construction documentation, progress reports, statutory and client approvals to be obtained, minimum local content

to be utilized, environmental restrictions on construction work, and quality management standards.

The end-product and incidental requirements are collectively referred to as the project scope.

- Organization of the project team roles by allocation of individual responsibility, ensuring that the complete scope is covered.
- For each team member or discipline, preparation of an operating plan outlining how the individual part of the project scope will be achieved. Arising from these plans, an integrated plan and a breakdown of the project work into defined packages sufficiently small for management purposes (discussed below).
- The project budget, broken down into elements identical to or compatible with the work breakdown structure.
- The project time schedule, in similar detail. The development of this schedule requires the development of the work breakdown structure into a network of sequenced activities.
- A system for controlling the project; that is, the means to be employed for measuring the technical quality of the product development, for measuring the cost and making cost projections, and for measuring progress, and comparing these with the planned values to alert management of need for corrective action. Usually, the control system will dictate the format and element size of the work breakdown structure, budget, and time schedule. Clearly, the control system and work element breakdown must facilitate progress reporting within a sufficiently short period to allow corrective action within an acceptable period, usually a week or two.
- A resource plan, identifying the human and other resources required for the project, and their source, timing, and cost, compatible with the project budget and schedule.

We will not elaborate further on the planning process, because this will be further developed in the discussion of the engineering work which follows.

3.3 Managing the project

The project team organization, referred to in item 2 of the plan, varies according to the size, complexity, and scope of the project. Smaller projects are best handled by a small team of all-rounders working full-time, rather than a multitude of part-time specialists: less communication and better teamwork more than compensate for any lack of more specialized skills, which can be obtained from a consultant if needed. Large projects can employ specialists to perform a single discipline of activities on a full-time basis. We will discuss the breakdown and performance of the work by discipline, according to the customary organization of a large project. This is not intended to imply that a single individual could not or should not perform a multiplicity of discipline functions.

The main groups of operational activities are engineering, procurement, construction, and commissioning.

Engineering includes:

- the design of the plant;
- specification of its component parts for procurement;
- preparation of technical documentation to guide and facilitate construction; and
- production of all other technical documentation needed.

The engineers also follow up on their work by provision of technical guidance, ensuring that the plant is procured, manufactured, and erected according to the design requirements.

Procurement includes the work of:

- finding suppliers or contractors willing and able to provide the goods and services required for the project;
- preparing commercial bid documents in collaboration with the engineers who prepare the technical specifications;
- getting competitive offers; and
- managing the process of deciding on the best offer for each bid, and carrying out negotiations to finalize purchase agreements or contracts.

The responsibility of procurement then extends to include follow-up activities required to manage the performance and administration of orders and contracts – manufacturing surveillance and expediting, correspondence and records management, management of delivery to site, payment management, disputes resolution, and order or contract close-out.

Construction includes all the work needed to build the plant on site. This can be organized in a variety of ways. A typical large project scenario is to appoint a construction management team led by a resident construction manager (RCM), and appoint one or more contractors to carry out the work. The various forms of contract and their implications will be addressed later. Whatever the system by which the work is performed and the contractual arrangements organized, the issues which have to be addressed include:

- work planning, co-ordination, and expediting;
- materials management;
- labour management and labour relations;
- safety management;
- technical information management and technical problem resolution;
- work quality control and acceptance;
- contract administration;
- record-keeping;
- financial management;
- site security, access management, and asset control; and
- final handover and close-out.

Commissioning is the work of putting the completed plant into operation. To do this safely, the commissioning engineers must first check that the plant has been properly constructed and is fully functional. The work also includes:

- operational planning;
- acquiring and training all the human and other resources needed to bring the plant into operation and to maintain it;
- methodically bringing each part of the plant into operation, and setting up the control systems;
- solving any problems of plant operation or reliability; and
- performance testing, and whatever formalities may have been agreed to complete the capital investment stage and proceed to the operating life of the plant.

Finally, the work of the project needs management, including:

- strategic direction and co-ordination of all disciplines;
- monitoring of progress and expenditure compared with schedule and budget, and initiation of corrective action;
- on-going review and audit of technical acceptability and conformance to the approved workscope;
- financial accounting and expenditure control;
- change control; and
- relationship management with the client and external entities (including the issue of reports).

On a large project, the project manager will be supported in this work by a project controls team, including cost engineers and planners (a.k.a. schedulers), financial accountants, and a secretariat charged with overall document control.

Chapter 4

The Engineering Work and Its Management

4.1 Planning the engineering work

The stages of project engineering may be described as:

1. Define the objectives, i.e. the end product of the project.
2. Decide how the objectives will be achieved – what methods will be employed.
3. Plan in detail each item of work ('work package') and schedule the performance of the packages in a logical sequence.
4. Do the engineering work and check that it is correct.
5. Ensure that the engineering designs and specifications are acceptably implemented in plant procurement, manufacture, and construction.
6. Commission the plant and finalize all technical documentation needed for operation and maintenance.

Items 1, 2, and 3 define the engineering scope of work in progressively increasing detail. Item 3, the detailed planning, cannot be completed until the conceptual design of the plant has been finalized ('frozen'). Thus, unless the project starts with a frozen conceptual design, there will be some overlap between the planning stage and the following engineering stage, and this is usually the case. In the following narrative, as was stated when discussing project initiation, we will address some activities as new subjects, although in practice some or all of the work should have been carried out during the study which preceded the project.

Definition of the project objectives starts with the development of a clear understanding of:

- the performance of the plant (what must the plant produce, and from what input resources, when, and how?);
- the processes by which the plant will function;
- all the special or local circumstances which will affect the construction or function of the plant; and
- any other specific requirements of the client.

To make sure that this 'clear understanding' is correct, is comprehensive, and describes exactly 'what the customer wants', the obvious procedure is to write it down as concisely as possible, review it thoroughly with the plant owner and operator, and get it formally approved before proceeding with design work.

In addition to the essential requirements mentioned in the previous paragraph, the engineer is confronted with choices of methodology during the design development, for instance which design codes to use, if this has not been specified by the client. The potential impact of choice of methods and codes must be considered. Often, there is little to differentiate between the possibilities other than the engineer's familiarity, but a choice must be made: a design based on a hotch-potch of codes is a recipe for disaster.

Because the procedure of developing this 'clear understanding' and choice of design methodology determines the entire design and construction of the plant, it is obviously of prime importance and deserves corresponding priority and effort. The performance and constructional standards which define the plant with 'clear understanding', and the design methods to be employed, will be termed the 'plant (or project) design criteria'. The documents and models subsequently produced to facilitate purchase and construction (in accordance with the plant design criteria) will be termed the 'design documentation'.

For the development of the design criteria, two scenarios need to be considered (there are many hybrids in practice). Firstly, the client may not know in detail what he wants, or he may know what is required but not have fully thought it out and written it down. For this first case it is the project engineer's job to establish the design criteria in conjunction with the client, generally by a process of reviewing the choices available and making recommendations, and to get the client's approval. If he does not do a thorough job of this, he risks at least wasted effort and delay in the design process, when fully developed designs are rejected on conceptual grounds, or at worst a final plant which is not what the client wanted.

In the second scenario, the client defines what he wants in a detailed specification. This is a prerequisite for purchasing a plant on a lump sum basis; without a detailed specification, the price is meaningless. However, the project engineer must still carefully review the scope definition before commencing the design process, clarify any ambiguities, and where choices exist, try to resolve them, before expending detail design manhours. Some choices may have to be deferred to a later stage, when further information is available, but he should at least define the basis on which the choices will be made.

A detailed checklist for the plant design criteria can run to 50 pages or more and varies with the type of process; most organizations that deal with plant construction or design have their own checklist. The essentials of a checklist (a checklist checklist!) have been included as Appendix 2.

Before moving on from the design criteria to the detailed planning of engineering work, there are several project management systems and procedures which need to be in existence. These are simply standardized methods of working, which are indispensable in a team operation to promote efficiency and avoid omission, duplication, and confusion. The following are part of the basic framework.

- *A document control system*, including standard practices for numbering documents, their filing, the system of document approval and authorization of issue, distribution control, revision control, and archiving.
- *The procurement interface.* How will engineering co-ordinate with the procurement function? Or will procurement simply be the responsibility of the engineers? The answer to the questions determines the scope and format of much of the engineering work. Agreed systems and interfaces also need to be in place for the procurement-related functions of materials control, workshop inspection, quantity surveying, and logistics in general.
- *The construction interface.* Inevitably this starts with the engineering definition of what construction work is required, to what standards, and how the work will be packaged. This leads to a follow-up stage in which construction information is issued, mainly in the form of drawings, and technical problems are resolved (requiring a site interface).
- *The cost control system.* This requires an understanding of the level of control – the breakdown of expenditure items (purchased items and engineering and management activities) by which costs will be authorized, reported, and controlled. There inevitably follows a need

for a consistent numbering system by which each item is recognized, usually called the code of accounts.

- *The project time schedule control (or planning) system.* Generally the same remarks apply as for cost control. Numbering needs are less exacting than for cost control, because 'roll-ups' of like items are not applicable. Generally the numbering systems employed for equipment and documentation are adequate.
- *A quality assurance system* and its implementation on the project. This is already partly implied, as we have listed 'checking of the engineering work' as part of the work. However, this may be a timely reminder that the checking and verification of work needs to be organized into a cohesive whole which will ensure that no critical aspect is left to chance, and that the overall standards of surveillance are adequate.

Advancing to the detailed planning of engineering work, we begin by grouping work into different categories. Most engineering work activities are associated with documents,[1] which are the product of the individual activities. Verbal communication is seldom acceptable, so each activity culminates in a document which is the essence of the work done or value added. The change of status of a document, for example from 'drawn' to 'checked' to 'issued for construction', is likewise associated with an amount of work and progress. Most engineering work is therefore conveniently identified for planning and control purposes by the associated document and its status. To quantify the work and break it down into manageable elements, it is only necessary to produce a list of documents, principally:

- design criteria;
- work plans, including work breakdown and work schedules;
- drawings (including vendor and third-party drawings to be reviewed and approved) and models;
- calculations;
- specifications and data sheets;
- procurement documents (requisitions, bid analyses, negotiation protocols, etc.);
- technical schedules, including the need for their ongoing update (equipment lists, cable schedules, etc.);

[1] A 'document' is a piece of recorded in formation of any type (hard copy, electronic, physical model). This subject is further developed in Chapter 20, Traditional Documentation Control.

- reports (technical, progress, inspection, site visit, meeting, etc.); and
- plant operation and maintenance manuals to be produced, and spare parts data.

Work not usually associated with specific documents may include:

- planned communication activities;[2]
- general management administration duties (work planning and work control schedule updates, informal communication, personnel administration, etc.); and
- follow-up of technical issues (including pro-active surveillance and response to questions and requests for assistance).

It is also possible to categorize many of these activities by associated documents, and to require documents as a report of the activity. To some degree such documents are necessary. For instance, a meeting requires a report. An instruction arising from a follow-up activity should be in writing, and the discovery of a serious discrepancy or failure should clearly trigger a management report, as well as corrective action. However, if this process is taken too far in relation to the real need for precise record, the excess of bureaucracy will become apparent to all, with the usual negative effect on attitudes.

For document-related work, engineering manhours are estimated by allocating budget hours to each document. For this purpose each document type is usually divided into a few sizes, each having a standard budget. Drawings are grouped by subject and size (A0, A1, etc.), procurement documentation by value and complexity, etc.

Non-document-related work may be estimated by adding a percentage to the document-related work, or by allowing a fixed time allowance per week, or by allowing a time package per followed-up item, for example purchase order. In the latter case, the final overall plan must be verified to ensure that adequate resources are available for reactive follow-up, in other words that there are enough engineers who serve by standing and waiting – in plan, anyway. Some unplanned work will inevitably materialize, and need to be done immediately. When no work demands arise, additional checking and surveillance activities may usefully be performed.

Sequence planning or scheduling of engineering work can be quite simple or very complex, depending on the project. For a simple project, where there is enough time to carry out activities in a logical sequence, and to commence individual activities when the required input information

---

[2] This subject is discussed in Chapter 25, Communication.

is available, the sequence may be as follows (obviously the type of plant and scope of project may change the picture).

1. Design criteria are prepared, and work is planned as outlined above.
2. The basic process design work is developed, culminating in process flowsheets and process data sheets for plant equipment items.
3. As process equipment data sheets become available, equipment specifications and procurement documents are written, enquiries are issued, and equipment is selected and purchased.
4. As purchased equipment information becomes available (including outline drawings, and access and maintenance needs) the plant layout drawings are developed (these are conceptual drawings with principal dimensions only) and maybe plant models are built. In this process, designs of the materials transport systems and plant structures and enclosures are developed.
5. Control philosophy and P&I diagrams are prepared. (This work may also be done in parallel with, or ahead of, the layout development, depending on the relative dependence on materials transport system design).
6. Based on the conceptual layout drawings and selected equipment masses and dynamic loads, equipment support and access steelwork, foundations, and enclosures are designed.
7. Also, as purchased equipment information becomes available, the electrical power supply system and switchgear are designed and procured, and plant instrumentation and control systems are finalized and procured.
8. The conceptual layout drawings are upgraded to general arrangement drawings, which reflect the detail steelwork and civil designs as well as actual equipment dimensions, and are dimensioned to co-ordinate interfaces.
9. Detail designs are prepared for vessels, chutes and other platework, piping and ducting, and electrical and instrumentation equipment and cables.
10. Technical packages are prepared for purchase of bulk materials and plant construction, including work description, bills of quantities, drawing lists, and construction specifications.
11. Technical documentation is prepared for plant commissioning, operation, and maintenance (including purchase of spare parts).

The corresponding flowchart is presented as Fig. 4.1. In this figure, 'design' has been used to describe activities whose main outputs are descriptions, calculations, and item sketches, whereas 'draw' describes

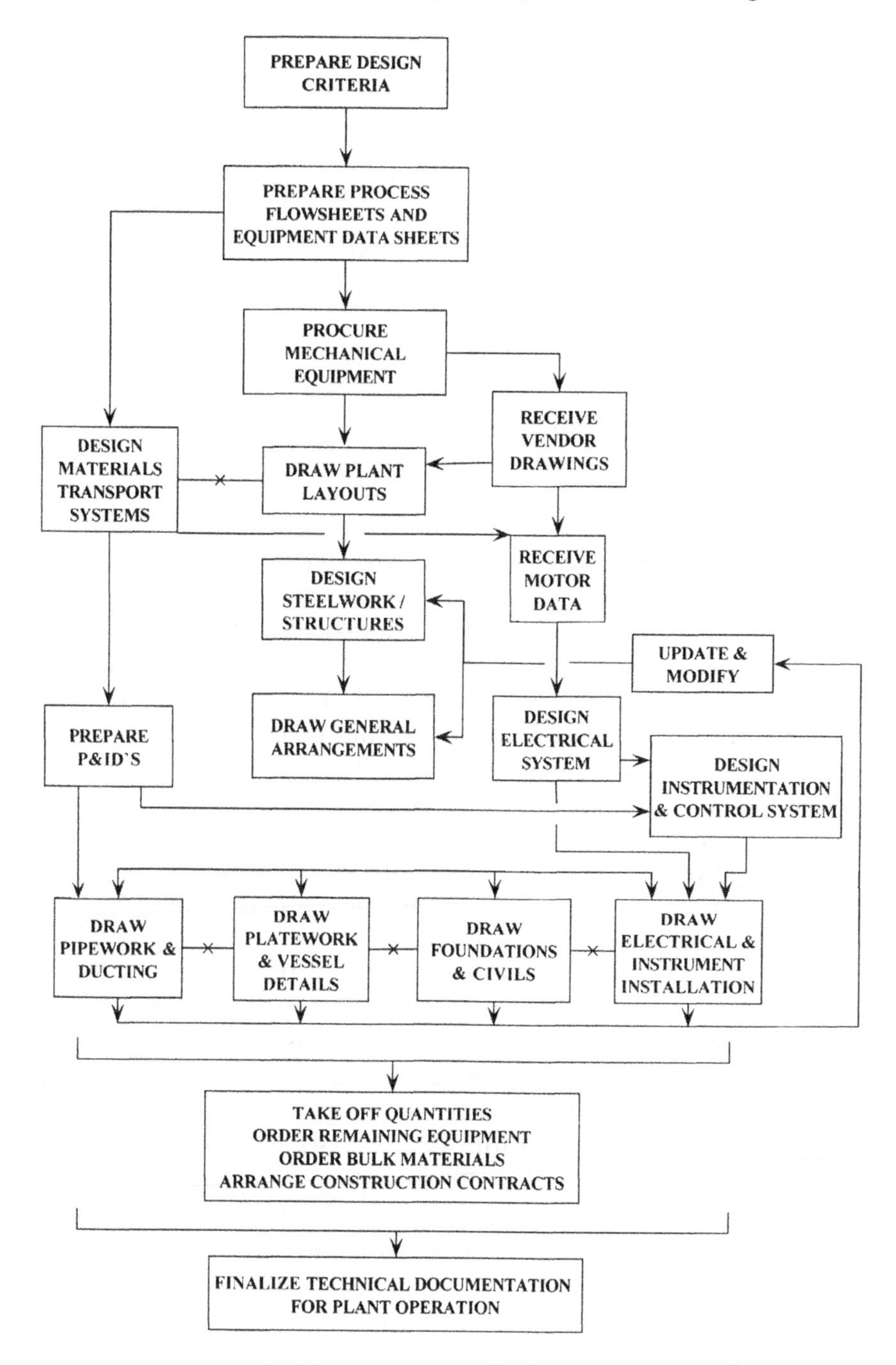

Fig. 4.1 Simplified engineering work sequence

activities which are centred on the arrangement of components within space, and the outputs are drawings or models. This distinction is artificial and is not so clear-cut in practice, nor can it be.

Usually, front-end engineering activities are performed in a preliminary pre-project or study stage, in which the conceptual designs and project baseline are developed. In the initial stage, no procurement commitments can be made, and the equipment data are not final. This brings about a separation of activities, which is presented as Fig. 4.2.

If the project baseline is not fully developed in the study stage (always the case, to some extent), then the project work sequence is something of a hybrid between Figs 4.1 and 4.2. In the most common hybrid, the final procurement of certain equipment items that are recognized as being critical to the schedule will proceed quickly after project authorization, while less critical items are deferred until the overall conceptual design and baseline have been reworked as in Fig. 4.2.

Even in these simplified examples, where activities follow a logical sequence of information flow, there is an amount of iteration. The layout drawings have to be available for structural design, but they are revised as a consequence of the design. When the connecting chutes, pipework, cables, etc. are added to the design, there are liable to be problems in fitting them in and supporting them, and plant access or maintainability may be compromised. Thus some revisions to the steelwork design and layout will usually be necessary. And so on for most activities, but the iterations are kept to a minimum.

The limitations of the project schedule, and the drive for reduction of project engineering and management manhours, seldom permit a completely sequential information flow. Usually, the design schedule has to be shortened. We should remember that the design process is not an end in itself – it is the servant of the project. Its various outputs have to be available according to a schedule that is co-ordinated with the procurement and construction needs. As a result there are demands for the early ordering of long-delivery equipment, the commencement of site construction work before design work is completed, and the ordering of bulk materials such as piping components before the design is completed (and sometimes when detailed design has hardly started). Simply putting more people on the job or working longer hours is not necessarily a practical method of acceleration, it is liable to cause a reduction in efficiency and increased errors. Besides, the schedule may be dependent on receipt of design information from relatively inflexible third parties. In practice the two most common devices to shorten the schedule are parallel working and assumption

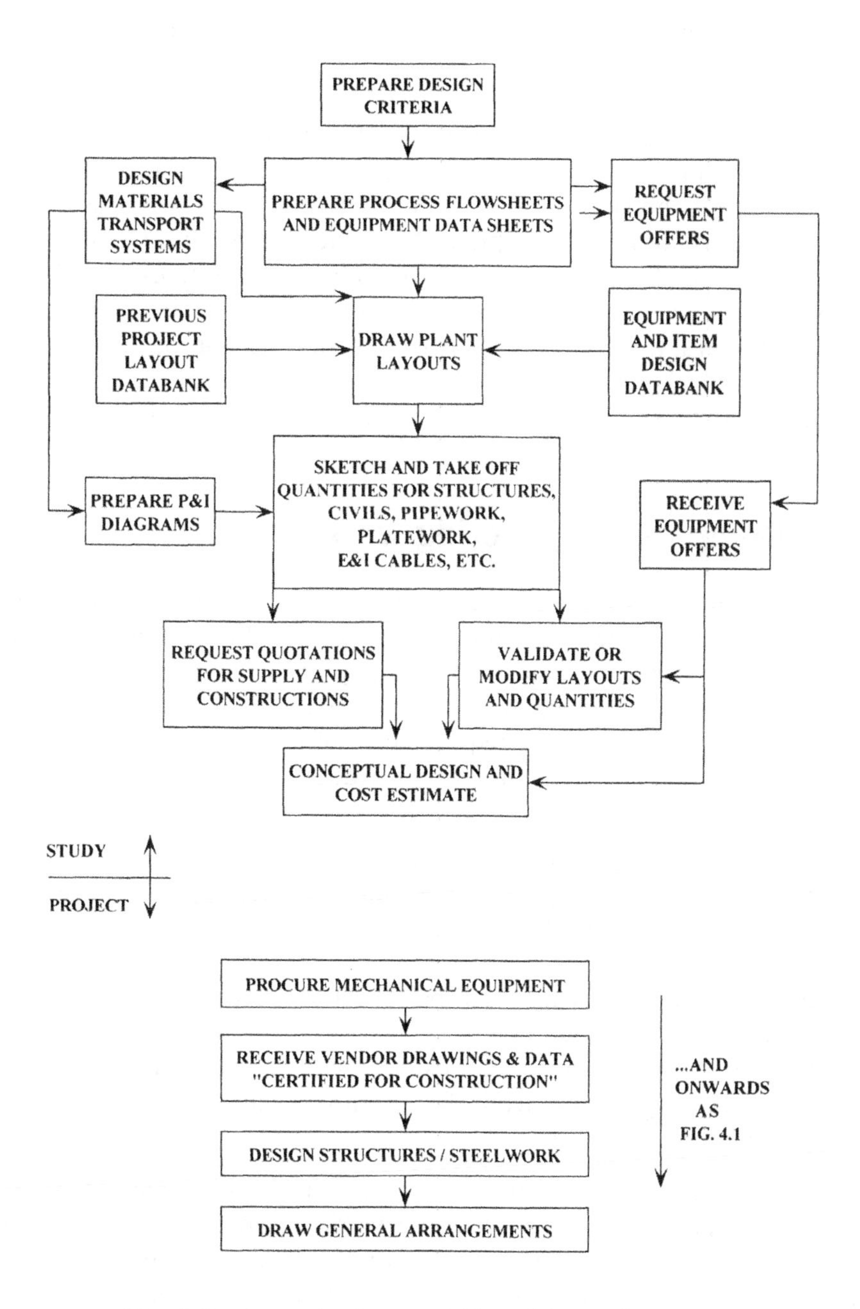

Fig. 4.2 Conceptual stage followed by project stage

of design information. In their application the two devices are almost identical.[3]

In conclusion, it should be apparent that a flowsheet or activity network can only provide an indication of workflow. It cannot provide a rigid definition of what tasks can be performed at what stage. There is an inherent flexibility to commence tasks without 100 per cent of the required information – at a price. It is the use of this facility, and the resulting challenges in terms of the judgement to be exercised and the teamwork needed to retain control of the work, that are some of the major differentiators of performance. The flexible approach is one of the factors which separate the 'can't-do' plodders from the achievers, although being too ambitious in this regard can result in chaos.

4.2 Doing the engineering work

In this section, we will review the management of the engineering work which has been planned, and enlarge on the execution of certain of the activities. It is not the intent to focus on specific engineering discipline methodologies and practice: we are concerned with the adaption of the discipline performance to the requirements of the process plant project.

Engineering management functions include:

- planning the work;
- monitoring the performance of the work (scope, quality, cost, time), comparing it with the plan, and taking corrective action as needed;
- co-ordinating the work of different disciplines and resolving any problems arising;
- managing changes (which has special relevance to co-ordination);
- managing resources; and
- managing engineering relationships with parties and functions outside the engineering team, for example client, procurement, and construction.

The management techniques of monitoring the performance of the work are dependent on how well the work has been planned, in the fullest sense of the word. The plan must include all elements required to ensure that all project objectives are met. All innovation and conceptual

[3] The making and control of assumptions, in order to expedite engineering, is one of the key issues of project engineering in a competitive environment. This is discussed in Chapter 27, Fast-Track Projects.

development should be completed at the planning stage, and, if not, a major crisis should be acknowledged. The successfully planned project should aim to be deadly dull in execution, a steady turning of the information development handle, with no surprises emerging.

4.3 Managing schedule, cost, and quality

There are many software packages and manual systems available to give accurate feedback on schedule performance, provided that adequate planning has been done and accurate reporting on task completion is available. But this feedback, although necessary, is historical; the art of engineering schedule management depends on anticipating the 'glitches'. Given that engineering at this stage of the project is fundamentally a job of information processing, the greatest care must be exercised in looking ahead to the next few weeks' information outputs, and understanding what inputs will be needed, especially between disciplines and across external interfaces such as procurement.

As is the case of schedule management, there are any number of software packages and systems to provide management reporting for cost control, and since this is regarded as a discipline in itself the subject will not be pursued in depth. The essence of any system is the breakdown of the project expenditure into a sufficient and appropriate number of components, which are tailored to enable the manager to get information on cost trends in time to take corrective action.

Engineering plays the major part in determining the cost of the project, firstly by determining the design and specification, secondly by deciding on the technical acceptability of purchased goods and services, and thirdly by re-engineering and technical compromise when cost trends are unacceptable. In this process, it is seldom possible for any cost reporting system to give timely management cost information, if the system is wholly dependent on actual prices and costs from suppliers and contractors. By the time such information is available, the effect on schedule and other consequences of modifying the design is likely to be severe or unacceptable. Therefore good cost management in engineering requires an awareness of the cost consequences of every design feature, at the time that each item is designed. As part of this process, all designs should be promptly checked for conformance to the conceptual designs and quantities on which the project budget was based.

There is another aspect of engineering cost management, which is the cost of the engineering itself. The control system is the same as for

Fig. 4.3 'Engineering plays the major role in determining the cost of the project ...'

purchased items, namely, the engineering activities are monitored against an itemized budget (usually of manhours). However, as engineering decisions and designs impact the direct field costs, there may be opportunities to reduce the overall project cost by increased expenditure on engineering, or, conversely, to save on engineering by spending a little extra money on equipment suppliers or construction contractors that then need less engineering input.

Last but not least of the management fundamentals is engineering quality management.

In the following chapters on the project environment, we will see that there is no absolute dictate on the quantity of engineering work that goes into a plant. For instance, the quality of work can always be improved by carrying out more checks and re-work, but this is not necessarily justified when any errors can be readily fixed up in the workshop or on site. There are some activities which are usually excluded from any quality compromise, including those related to the plant's structural integrity and safety of operation. So before planning the quality management work, the underlying quality policy issues need to be clearly exposed and agreed, for which purpose the consequences of possible engineering shortcomings need to be evaluated. Actions necessary to ensure the correct engineering of critical items may then be planned. For non-critical items, economically based quality assurance procedures may be drawn up, and a reasonable contingency may be included for remedial action in the field or workshop.

The detailed planning of engineering quality management is a matter of procedures: deciding what checks, reviews, and audits will be made, how they will be made, and when they will be made. The quality system may have to comply with the requirements of standards such as the ISO9000 series, or the plant owner's stipulations. A few general principles to bear in mind are the following.

- The quality of engineering work is generally assured by the acceptance of responsibility by individuals, who sign their names to a document. Process plant engineering is a team effort, therefore the signatures may not reflect total responsibility for the work represented on the document; rather each signature reflects that certain activities impacting the document have been performed correctly. The quality management process is undermined unless it is clear what those activities are and how they must be performed. Procedures, job descriptions, and checklists have to make it clear. Ideally, when a document is signed as checked or reviewed, the checker/reviewer should add the reference number of the checklist or procedure that he has followed. As a general rule, quality assurance depends on the correct definition and understanding of responsibility.
- An audit trail for design information, detailing the source document and its revision status/date, is an indispensable quality tool. This can also be effective when used in reverse in controlling design changes, when the omission to revise all consequential changes is a frequent problem.
- Because of revisions – engineering is an iterative process – a signature always needs a corresponding date.

It may be noted that all of the comments on engineering quality relate to ensuring that the planned objectives are met – none relate to 'engineering excellence'. This should not be taken to imply that engineering is just a matter of doing what has been planned – excellent engineers have plenty of scope for 'making the difference' in the conceptual stage, for developing the most effective design details and, all too frequently, for producing solutions to unanticipated problems. The subject presently discussed is quality management rather than quality itself, and it needs to be stressed that even with a team of brilliant engineers (especially in this case, possibly), quality lies not just in individual brilliance, but probably more in meticulous attention to detail.

4.4 Co-ordinating engineering work

In general a process plant is engineered by a multi-disciplinary team, and the work is further divided into drawing-related activities and other engineering activities such as design calculations, specifications, and equipment purchase. The work of the various engineering disciplines and their members has to be co-ordinated in terms of:

- function (for example, gravity flows require appropriate elevation differences, the electrical supply and motors have to be matched to the driven machinery, etc.);
- structural loading (which is greatly dependent on process and equipment information);
- the occupation of space, including physical interfaces and the provision of access for construction, plant operation, and maintenance; and
- the sequence and timing of information flow.

Each discipline has to check and approve the documentation of others, where there is an interface or possible clash due to occupation of the same plant-space. There is no substitute for this process if the individual accountability of each discipline is to be maintained.

In our discussion of planning and schedule management, we discussed how individual activities were tied together by the logic of information flow. We noted that, even for the sequential performance of engineering, there were inevitable iterations as different disciplines' work impacted on one another, and that more such iterations were needed for the usual circumstances when sequential working is compromised to shorten the schedule.

Engineering work co-ordination centres on improving the inter-discipline information flow in order to shorten the periods of iterations and reduce their number, thereby shortening the schedule (while maintaining the quality of the end-product). It aims to mitigate the consequences of out-of-sequence design development.

As for other activities, efficient co-ordination starts with a good plan. The plan should be drawn up on a needs-driven basis, and include a list of discipline interface information requirements, specifying what is required, and by what date, and who is responsible. It is also necessary to draw up a matrix of review-and-approval requirements, specifying type of engineering document on one axis, and the disciplines concerned on the other.[4]

[4] This is developed further in Chapter 25, Communication.

Careful review of the initial plan by the design team leaders shows up where special steps need to be taken to reduce iterations or information delays, and to modify the plan accordingly. Such steps typically include the following.

- Adoption of conservative design parameters, which will not usually be changed if subsequently proved to be over-conservative. Examples are: allow greater clearances within a structure, allow for higher structural loads, and allow for higher driven machine power requirements. Clearly the benefits of such actions have to be balanced against the cost implications.
- More conceptual work to eliminate uncertainties, which may otherwise result in design reiteration.
- Changing the design to a more standard or better-proven design. This may include the purchase of a proprietary item or package, in place of a customized project design.
- Looking for previous similar project experience, or a suitably experienced consultant.
- Where uncertainties arise out of future procurement decisions, changing the procurement policy. For example, by negotiation with a selected supplier rather than competitive purchase, or, less radically, by specification of design features which otherwise would be determined by competition. The cost and commercial implications obviously have to be evaluated.

During implementation of the engineering work, a similar range of remedial steps may be considered if it becomes necessary to bring the work back on schedule.

The greatest challenge to co-ordination usually lies in the drawing office. The challenge is best illustrated by some of the problems to be avoided.

- Drawings chronically late for construction needs while design problems are resolved.
- Support structures which do not fit the supported equipment.
- Clashes between piping and steelwork.
- Items which cannot be erected on site without being cut into pieces.
- Process equipment at the wrong elevation for flow of process materials.
- Inadequate plant access and maintainability.

In order to identify the actions needed to avoid such problems, we will consider the causes.

- The first and probably most frequent cause lies in the *drafting process,* or process by which the plant is modelled. The possible ingredients here are drawings which have inherent errors, drawings which are difficult to interpret, and drawings which are inadequately reviewed or checked by other disciplines, and therefore clash with other disciplines' or overall needs. Apart from the obvious remedies of more skilful and better-trained people producing and reviewing the drawings, and more thorough co-ordination procedures, there are many systems and software available to improve drafting co-ordination and the visualization of the design. These will be discussed in Chapters 10 and 28.

- A fundamental cause, which often underlies drafting problems, is *inadequate initial engineering* or forethought. Error and iteration are reduced, and the burden on the draughtsman is reduced, if drawings and the inherent problems to be overcome are thought out clearly by all the disciplines involved, at the conceptual design stage and before detail design commences. An essential ingredient of this forethought is a library of design standards and standard designs for common plant items, which will be considered at some length in Chapter 19.
- Process plants are configured around process equipment, which is for the most part built to *proprietary ('vendor') designs.* The quality and timing of receipt of the vendor drawings, and of information such as structural loadings and electric motor sizes ('vendor data'), are critical to the plant design effort. The provision of adequate and timely vendor information must be a prime objective of equipment procurement procedures, by specification, commercial terms, and by follow-up.
- Finally, *changes can disrupt the design process* far more than is expected by the people initiating the changes, resulting in errors and delays. Some changes are of course inevitable, for instance if the plant functionality or safety is found to be questionable. Change management requires careful attention in order to implement accepted changes, to be able to argue against changes whose real cost is unacceptable, and perhaps most important of all, to stop unapproved and therefore unco-ordinated changes from being made.

4.5 Management of engineering resources

In our review of the planning of engineering work, we addressed the quantification of manhours and the sequencing of work without really addressing the availability of resources, that is, the human resources, the tools such as computers, and the facilities needed. From the detailed work breakdown, which includes the resource type and hours needed for each activity (or document), and from the required work completion schedule, it is a simple matter to plot out the input hours required of each resource category against time. There will inevitably be periods of unacceptably high or low resource utilization, which are incompatible with reasonable standards of work continuity, team-building, and resource availability. The schedule usually has to be 'smoothed' by a process of compromise to yield a schedule which can be resourced with reasonable confidence and is considered to be optimal. The process of compromise is made more difficult and challenging by the fact that the schedule sequence logic is also subject to compromise if necessary, as discussed previously, and in fact both aspects must be considered together.

The management of process plant engineering resource costs and schedule durations is frequently problematical and occasionally disastrous. Often the problems stem from a technical error, or series of errors, which are discovered too late and require remedial action. Equally often, however, the problems stem from management failure, that is, from failure to plan the work and control its execution according to the plan. The plan may be unachievable because the project team is not sufficiently competent to meet the challenge, or because irresistible (or insufficiently resisted) external factors dictated an over-optimistic commitment.

Really disastrous overruns may occur if not only the work performance but also the monitoring of progress against the plan is faulty, so that the project manager has insufficient time to take appropriate measures. This can be caused by simply leaving items of work off the plan; usually the quality-related items such as layout reviews and corresponding re-work allowances. Flawed designs go undetected, and remedial work is deferred until the consequences of omission come home to roost. Another frequent cause of over-optimistic progress reporting is failure to measure accurately what work has been completed and what is outstanding, because documents are regarded as complete when they are subject to change (the consequences of which are routinely underestimated). The experienced project engineer learns to ensure personally that the plan and its monitoring retain their integrity in all these regards, because a third-party planner (or scheduler), however well trained, may have insufficient 'feel' for these and similar aspects.

Second Cycle
Environment

Chapter 5

The Project's Industrial Environment

5.1 The industry and the client

'The industry', which is our subject, is the industry of designing and building process plant. This is obviously peripheral to another industry, which is that of utilizing process plant in order to convert feedstock to more valuable products – the 'client' industry. We will begin by discussing some characteristics of the client industry, before going on to discuss the industry whose purpose is to satisfy the client's needs for new plant, for plant upgrades, and for associated services. In general we will discuss projects for the provision of complete process units, rather than other types of project and work which are essentially similar to, or a fragmentation of, the work of building a complete new unit or combinations of units. Our aim here is to identify some of the aspects of the industrial environment which have a major influence on the way that project work is done.

In the process plant industry, even for superficially similar applications, there are many plant design variations and client expectations to be considered. The first design rule is possibly to take nothing for granted about the needs of the client or the market, but to methodically examine the characteristics of each project or prospective project, starting with as much lateral vision as possible.

Considering the characteristics of the client and of the plant, we will note in particular the following aspects: feedstock, product and process (classified as discussed in Chapter 2), scale of operation, plant owners' and operators' behaviour, financial and strategic influences, and environmental considerations. Several of these parameters appear in fact to be interdependent, but they all need to be taken into account as part of the exercise of understanding what differentiating practices and expectations are associated with a particular field of project work. The generalizations that can be made are illustrative of attitudes and behaviour, and of course there are plenty of exceptions.

The hydrocarbon processing industry deals with feedstock and product which are minerals of moderate unit value, moderately hazardous (mainly due to fire and explosion risk), and are almost invariably processed in the fluid state. The attitudes of the industry seem to be influenced mainly by the vast quantities of materials and finance involved, and the inelastic market demand. Economies of scale for both production and distribution have led to an industry dominated by very large, vertically integrated global companies, and the process industry within this framework has tended to be organized accordingly, with highly standardized views on how a plant should be designed and how a project should be structured. From the American Petroleum Institute there is a formidable body of standards and standard practices, and most national standards institutes have many relevant standards and practices of their own.

The typical hydrocarbon processing client therefore expects a highly standardized approach to design, with emphasis on thoroughness and use of the most experienced equipment vendors rather than any innovation. This design approach greatly facilitates the management of projects in a well-organized way, well-controlled from start to finish, even when there are very tight time schedules. (The less organized project teams are still capable of making a mess of the job!)

The metallurgical industry does not enjoy a similar degree of standardization. The processed materials are rocks of varied form and substance at the front end of the plant, and the products are also solids. There are many very different types of ore to be handled, and a wide variety of processes and products. Solid-phase materials transportation means that plants and their layouts tend to be customized, with scope for innovative solutions. Most ores are mainly waste which would be expensive to ship, so at least the front end of the plant has to be built at the mine, and be designed to suit the topography. The processed materials are for the most part relatively non-hazardous. When, as in the case of the platinum industry, highly toxic processes are necessary for the refining of the product, the refinery is usually built and operated as a separate entity from the concentrator and smelter, requiring a different approach and different behaviour. For the majority of low-inherent-hazard processes, the lesser inherent risks facilitate

Fig. 5.1 Platinum group metal refining unit – closer to a petrochemical plant than to an ore processing plant

innovation without compromising safety. (The less careful project teams are still capable of designing a dangerous plant!)

The characteristics of the chemical plant industry tend to lie between those of hydrocarbon processing and metallurgical extraction, being more similar to hydrocarbon processing plant when the feedstock is fluid and hazardous, and more similar to metallurgical plant when the feedstock is solid and non-hazardous.

There are of course many types of plant which process materials that are toxic, such as poisonous or radioactive substances, and corrosive fluids. Such types of plant invariably have a technology of their own, developed and proven to be adequate for the hazards involved, and any deviation from established plant designs or standard operating practices requires careful consideration, and possibly third-party or regulatory examination and approval.

Plants for processing water, either to produce potable water or to treat effluent, generally involve neither hazard nor the transport of solids other than sewage or minor detritus in slurry form (the most transportable slurry of all). The main emphasis when building such plants is on reducing capital costs to a minimum, while remaining (just) within whatever standards of product quality and construction are specified. Innovation by technology advance (however minor), by structural optimization, or by contracting method, is at a premium, with minimal down-side potential (but some contractors still lose their shirts!).

Power plants, for the generation of electricity, are not normally included under the definition of process plant. Certainly, when the prime mover is a diesel engine, water turbine, or simple-cycle gas turbine, the plant's characteristics more closely resemble those of a single piece of equipment than those of a process unit. However, a major coal-fired power station, and the project to build it, have much in common with process plant work: the system is built around a flowsheet with many participating items of equipment and connecting pipework and materials handling devices. In this case, the most important differentiating characteristic is that the product – the kWh – is of relatively low individual value, is required in massive and predictable quantities over the future life of the plant, and cannot economically be stored. As a consequence, there is a major emphasis on overall energy efficiency and the factors which contribute towards it, and usually a willingness to pay more in capital to achieve this goal. Capital is relatively cheaply available for the perceived low risks involved. These factors also contribute to a relative willingness to accept longer project completion schedules, provided that some cost reduction is thereby obtained, and also assuming that generation capacity growth has kept up with the demand.

The comments above are of course highly superficial in relation to industries which have developed technologies that people spend their careers in acquiring. Our purpose here is not to discuss technology, but its project application, and the main point to be made is that each type of plant industry has developed in a way that has been driven by characteristics related to the nature of the process and processed materials and the scale of operation. The consequences of this development are not restricted to technology, which may be presented in a package, but extend to the attitudes and general behaviour of the people involved in the client industry, in particular, their attitudes to innovation, work standards, and cost reduction. These attitudes need to be understood for successful project work. It also needs to be understood that there are many anomalies; for example, petroleum companies that have acquired mining interests may attempt to organize metallurgical plant projects with the same attitudes as for hydrocarbon processing plant, often to the detriment of the project and the contractors involved.

5.2 The client and the project management

The relationship between the client and the project manager and project team is likely to have a major influence on the way in which the project is managed. The project manager may be a senior executive within the client organization, and he may have total authority to fulfil the mission up to the point of handover to an operating entity. Or he may be a relatively junior member of a large organization, with imited authority to co-ordinate the work of specialist departments. The process plant owner may directly employ the project manager and team, or the owner may buy a turnkey plant for a lump sum from an independent contractor, or he may employ a company as managing contractor, headed by a project manager, to design the plant and manage its construction. And of course there are many variations and combinations between these cases. In order to make progress with the analysis of the relationship between the client and the project manager, let us consider two extremes: the turnkey plant option, and in-house management by a senior executive of the owner's organization.

For a turnkey plant to be built, there has to be first a process of plant specification and competitive bidding, to establish who is the contractor, what will be delivered, when, and at what price.[1] Thus at the outset of the turnkey plant project, the project objectives ought to have been clearly set out, with little room for cost increase to the client, or for uncertainty about the scope of supply.

In the case of in-house management, and especially so when the responsible executive is very senior in the company hierarchy, he can be expected to ensure that the project will be seen to be successful, and in particular be completed within the authorized budget. There are many ways for the executive to do this, including the setting of targets that are easy to meet, without the problem of competition. As corporate financial controls normally require a fixed project budget, one of the methods more often employed to ensure the financial success of the project is to allow some flexibility in the scope of work. For instance, the executive may decide, when the project is at an advanced stage and cost trends are clear, whether to provide a fully equipped plant maintenance workshop,

---

[1] There are a number of variations to this theme, which will be explored in Chapter 7, The Contracting Environment.

or whether to save the associated capital cost increase and rely on external contract maintenance. In other words, shift the cost of this service from capital cost to operating cost. Similar scope variations are possible with many nice-to-have items, such as standby equipment or maintenance-saving features. If these practices are permitted, the cost estimate and cost budget for the plant become simply self-fulfilling prophesies, and the pressures of budget compliance are not nearly as great as in a turnkey environment.

Aspects of such management practices are likely to appear in all vertical relationships in the project hierarchy, from the chief executive of the corporation that owns the plant down through the project team and sub-contractors. A responsible party tends to hold those reporting or sub-contracting to him accountable in detail, while attempting to preserve the maximum of flexibility of performance target for himself.[2]

In the following pages we will in general assume that there is a competitive environment, and that projects have to be completed in accordance with technical and scope specifications, on time and within budget. Those project practitioners who operate in less demanding circumstances should quietly enjoy their good fortune.

5.3 The process plant project industry: the 'indirect cost' of a plant

A process plant project embraces several distinct types of work, which are customarily carried out by different industrial groupings. Work components include:

- the underlying process technology, briefly addressed in Chapter 2;
- the design and manufacture of plant equipment items;
- the manufacture of bulk components such as structural steel, pipes, and electric cables;
- structural steelwork, piping, and vessel fabrication in accordance with project-specific drawings; and
- site construction to bring the components together into a working plant.

These individual project work components have to be defined, procured, co-ordinated, and controlled by a project engineering and management effort, which is the subject of this book.

[2] This aspect of human nature bedevils the use of performance statistics for project risk analysis, and for the setting of contingency allowances for project budget and schedule. The subject is further discussed in Chapter 9, Studies and Proposals.

To understand those facets of the plant design and construction industry which most affect the lives of people who design and manage such works, it is illuminating to start by asking the question 'Why do we need such people at all?' The question can usefully be answered by dividing their work into two parts. There is a minimum engineering and management portion without which the project cannot be executed at all. This work will be more in plants of original design, and less where previous designs can be utilized, but the work cannot be dispensed with altogether, otherwise there will be no instructions to purchase or build, no basis to set out the site, and no plant configuration. This part of the engineering and management work is almost like any of the plant physical components, an indispensable part of the plant. The remaining engineering and management work inputs are characterized by 'added value', that is to say that their execution and cost have to be justified by comparison to the added benefits or reduced overall costs arising from each item of work.

As an example, it is possible to get the plant constructed by simply appointing a chosen contractor to do the construction work at whatever hourly rates may be agreed by negotiation. But it is usually more effective to spend money on the engineering and management work of drawing up a specification and contract for the construction work, and soliciting competitive bids, thereby gaining the services of a more target-orientated and less costly construction company. The incremental engineering and management costs are lower than the value of the resulting benefits.

We will examine the quantitative relationship between engineering and management work (broadly described as indirect costs) and hardware and construction costs (direct costs) in Chapter 8; the present discussion is restricted to qualitative aspects. The division of project costs into direct and indirect components is widely practised, and has very important implications, but it is often little understood. What seems to escape many is that it is a purely artificial distinction.

There are clients who value the outputs which incur direct costs – in particular, hardware – far more than the outputs which incur indirect costs, because the direct cost items are tangible. In the value system of these clients, a project with a low ratio of indirect costs is evidence of efficiency and good value for money. In fact, if any single item of 'direct cost' is examined, it can be seen itself to be separable into a direct and indirect cost component. For instance, a contract to provide pipework may be seen (and reported) as a single direct cost item. As far as the piping contractor is concerned, however, there are direct costs such as the materials purchased and the labour hired for the job, and indirects such as the costs of his own detailed engineering for the job and his own contract management and overheads. In the case of a piece of equipment such as a pump, the supplier's costs may be broken down into the directs of labour, material, and component costs on the one hand, and the indirects of customized engineering, sales, factory overheads, and order management costs on the other. The 'direct' costs can be further broken down by investigating sub-contractors and component suppliers, *ad infinitum*!

By carrying out such further breakdown, it is easy to reduce the average plant project direct cost proportion to below 50 per cent, with scope for even further reduction for those interested in futile intellectual exercises.

Defining what may be classed as an indirect cost is not just a question of terminology. The actual work performance of many indirect cost items may be moved between the project engineering and management team and the fabrication and construction contractors, depending on how the contractors' workscopes are defined. In particular, this can be done by purchasing either large, all-inclusive packages, or many small items.

More cost-effective plant design may make certain pieces of equipment redundant, leading to reduced direct costs as a result of improved performance in the indirect cost sector. The reverse process, of performing substandard engineering and management work and causing an increase of direct costs, is also a frequent occurrence!

For many clients, and indeed for many project practitioners, the problems of distinction between indirect and direct costs discussed above are academic and irrelevant. To these people, the understanding of what work is to be done for a given project, and how it is to be done, is clear and unvarying. Any questioning on how the work can be broken down between direct and indirect, and what constitutes an appropriate proportion of indirect costs, is likely to be taken as an attempt to disguise inefficiency. ('We *know* that 16 per cent indirects is the right figure for this type of plant.') The issues surrounding indirect cost ratio will be further debated at some length in the later pages, as they have an important influence on how work is done. We will hope to convince the reader that the reality is not as simple as the 16-per-centers would have it, and that those who are prepared to acknowledge and study the reality have scope for better performance than their more myopic competitors.

Chapter 6

The Commercial Environment

6.1 Principles of procurement and contract

The cost of engineering is a relatively small part of the cost of a process plant, say 5–15 per cent, depending not only on the complexity of the plant but also on what activities are included within 'engineering' or are included in procured item packages and construction contracts. Most of the cost of the plant, the direct field cost or DFC, is expended on purchasing equipment, materials, and construction. In designing the plant, a lot of effort and skill goes into producing economic and cost-effective designs which are aimed at minimizing the DFC while maintaining functionality and technical standards. Just as much effort and skill need to go into maintaining the interface with suppliers and contractors in the best way, commercially. Arguably, more money is wasted, and more opportunities are lost, by poor technical/commercial interface than by poor design.

The 'interface' with a supplier or contractor is governed by a purchase order or a contract, respectively. Although purchase orders and contracts have differences in format and content, they are essentially the same, in that they are both a form of contract which is arrived at by a process of offer and acceptance. The two-sided nature of a contract is often insufficiently understood by engineers – both sides have obligations and can expect to be penalized for non-performance. Even the fundamental process of offer and acceptance is subject to reversal. If an offer is accepted subject to certain conditions or changes, the conditional acceptance effectively becomes a counter-offer and the roles are reversed. In fact as suppliers' bids are seldom accepted without qualification, the purchase order or contract issued by the project manager is usually the final offer, and is not binding unless accepted by the supplier or contractor.

The fundamental principle of procurement in a market economy is of course that the best prices are obtained by competition. Competition can be effected in several ways, the most usual (which we will assume as a default case) being that the goods and services to be procured are defined, and competitive bids are solicited.

Alternatively, competitive circumstances can be introduced indirectly, by:

- negotiating with a single supplier, under threat of going to competitive bid if an acceptable deal is not offered;
- agreeing on a price which has already been established by the market; or
- a process of genuine[1] partnership in which participants, who would otherwise relate as suppliers or sub-contractors, share in the overall competitive forces and profitability to which the end product, the process plant, is subject.

However, the development of a commercial relationship or business deals by any means other than fair and open competition is susceptible to corruption, or even the perception of corruption, which can be equally damaging.

One of the basic problems confronting the project manager, since the beginning of the history of projects, has been that of maintaining commercial integrity, to which there are several aspects. The concepts of commercial integrity are based on the cultural values of a particular society and are therefore variable in place and time. The project manager and his team are responsible and accountable for spending large sums of money, and need to understand what practices are expected and what are acceptable in the circumstances, at the risk of being considered guilty of corrupt practice (assuming that is not the project manager's actual intention!). Procurement procedures therefore tend to be elaborate and subject to external audit and supervision. This is as much in the project team's defence as it may be a burden, given the tendencies of even some of the most unexpected people and organizations to line their pockets or employ corrupt practices when the opportunity arises. These procedures need to follow the needs of the country and the client, considering that what is regarded as corrupt practice in one society may be acceptable behaviour in another.

How the engineering–commercial interfaces are set up is evidently critical for the planning of the project. The interfaces effectively define

[1] This adjective is intended to imply that sham partnership arrangements are common in practice.

the format of much of the outputs of engineering work, the documentation created for purchase. The content and timing of this documentation must reflect a balance between technical and commercial priorities. For example, it is counter-productive to conduct lengthy engineering studies to minimize the steel tonnage of a structure, when as a result there is so little time for procurement that competitive prices are not obtainable without delaying the construction schedule. The financial gains of reduced mass are likely to be far outweighed by higher unit prices.

It is also counter-productive to technically format a construction bid in a way which is unacceptable to the target market. This can happen, for instance, if lump sums are invited for work which is regarded by the bidders as unquantifiable, which may result in high bid prices because the vendors include high contingencies, or because few bids are submitted. The engineering costs saved by the simple bid format may be much less than the increase in the direct construction costs.

Before finalizing the engineering plan, it is necessary to agree a procurement plan in which the commercial policies and procedures are recognized, and are given due weighting in relation to their implications. It is also vital to agree the format of purchase orders and contracts, and to decide on basic contracting relationships, such as whether specific items of construction work will be handled by lump-sum contracts, by rates-based contracts, by direct labour hire, or by the client's own resources. Such decisions (which will be discussed in Chapter 23) have fundamental implications, such as the way in which site work may be related in sequence to the finalization of design. Failure to do this planning properly – a frequent shortcoming – can result in ten times the amount saved on engineering being lost in procurement and construction.

For a competitive bidding process to be effective, not only must the format and content of the bids be appropriate to the market but also the goods and services purchased have to be accurately defined to make the competition meaningful and the contracts enforceable. Furthermore, the purchaser – and that mainly means the engineer – must fully understand his side of the two-way contract, and be ready to fulfil those obligations (in addition to payment) on time, or risk having to settle claims that may nullify the benefits of competition. The obligations in question may include the provision of information (usually drawings), approvals, materials, site access, or third-party activities.

The education of an engineer centres on the understanding and logical application of scientific principles and technology, and inculcates a set of values based on the inherent worth of what is produced. The thinking and behaviour which appears to be most successful in commercial dealings or negotiations is radically different, and this needs to be understood by the engineer. Some people or cultures seem to have this commercial ability by intuition. Others never understand it and it costs them dearly: they will never get the best bargain, and they may be 'robbed blind'. The basis of this ability is possibly manipulative thinking: the ability and the intention to influence people by means other than logical argument or force (both of which are readily understood by most engineers). This is supplemented by a mindset in which gain is made by smart trading, or negotiation, rather than by value added inherently – scheming, devious behaviour, or great commercial insight, depending on your point of view. Successful commercial negotiation skills are as essential as engineering skills to the outcome of the project, and much more important in some environments.

The art of negotiation can be learnt. In general, the objectives are to explore matters of common interest between the negotiating parties, identify the basis for agreements which can be beneficial to both parties, and reach actual agreements. Usually, there is an overlap between the minimum terms acceptable to one party and the maximum that the other party would be prepared to concede to reach an agreement. The art to be exercised by the individual negotiator is to recognize and quantify the range of overlap of possible agreement, and ensure that the eventual agreement reflects the best deal for his principals, within the achievable range.

More specifically, if you are buying something, the maximum price that you are prepared to pay (influenced by what you can afford, and what you can get from other sources) is often greater than the minimum price for which the vendor may be willing to sell. Your negotiating aim is to settle on the minimum price.[2] To do so, you need to know or guess what the vendor's minimum terms are, and you need to ensure that he does not know your maximum. You may manipulate him, for instance, by frightening him that he may lose the business altogether, and there are various obvious and subtle ways of doing this. The subtle methods are more effective, especially if the vendor does not realize that he is being manipulated. Keeping your own position secure is often a challenge, especially if some witless young engineer – who possibly

[2] Originally, the text here was '... and pocket the difference.' This has been amended, on review, to clarify that the author is not advocating the practice (common in certain environments) summarized by '5 per cent into my offshore bank account, and the order is yours'.

thinks technology is everything, and negotiating is unfair – passes too much information to the vendor.[3]

Like so many subjects touched on in this book, negotiation is an art in itself on which several excellent manuals exist. A more detailed discussion is beyond our scope, but it is relevant to discuss one aspect, which is that negotiation is ineffective unless the agreement reached is comprehensive of all aspects of the proposed co-operation, for example the full terms of the purchase order and the specification of the goods purchased. Little is likely to be gained if, when agreement is reached on what seems to be a bargain price, there remain other necessary aspects of agreement to be reached, such as technical details. Once the vendor has secured a commitment, there is clearly no further threat of losing the order. Ground lost in the preceding negotiation may well be made up in the addenda!

The 'price' of a proposed order or contract is not usually a clear-cut matter. The real cost to the purchaser may be affected by variations in terms of payment, provision for taxes and duties, terms of exchange rate variation, or any one of the myriads of conditions of purchase. The price may also be affected by variations in the scope of supply or in quality requirements. Quite frequently, a vendor will deliberately introduce some vagueness into his offer in order to leave room for negotiation. On the other side of the deal, purchasers may claim to have fixed tendering procedures which allow for no price changes or price negotiations after the submission of bids, but in practice they may effectively negate these stipulations. They may keep the price intact, but negotiate conditions which would affect the price, such as the terms of payment, the technical specification or scope of supply. In order to gain the full benefits of negotiation, all technical and commercial matters, including the handling of anticipated future changes and additions, must be settled before reaching final agreement.

Unless it is ordained in the persons of a separate procurement department, engineers lacking negotiation skills are well advised to ensure that they receive appropriate help in this regard, or they may well end up being totally outmanoeuvred by somebody who does not know one end of the plant from the other. It is also appropriate to note that as the relationship of contract is fundamentally a legal one and subject ultimately to legal interpretation, professional legal input is necessary in the drawing up and management of contracts. This may be limited to a consultancy role if sufficiently knowledgeable procurement professionals are employed.

---

[3] Possibly over cocktails ... and possibly while our young engineer is being told that he's 'the only real engineer in the office'. Manipulation is a two-way process!

Chapter 7

The Contracting Environment

7.1 Ways of building a plant

We will discuss purchase and contract, in so far as they are part of the project's execution, in more detail in subsequent cycles, concentrating on the engineering implications and interface. In what follows we will discuss the contractual aspects of setting up a project organization, namely the contract environment in which the project engineer himself must work.

To set the scene, we will first discuss the options available to an investor or potential investor, 'the client', in a process plant project. His fundamental choices are either to do the work with his own resources, 'in-house', or to get an engineering company or consultant to do the work, or at least to manage it.

Disadvantages of doing the work in-house include the following.

- Loss of the benefits of focus: most clients' core business is the manufacture and marketing of product. The engineering and management of major plant construction projects is often not considered to be a core activity.
- In the case of projects which are large in comparison with the client's in-house resources, all the problems of hire-and-fire to cope with uneven workload, or alternatively the debilitating effects of hire-and-don't-fire in such circumstances.
- Loss of competition, from which the in-house workforce is inevitably sheltered, potentially resulting in unimaginative, inefficient performance and sub-optimal designs based on individual preference (for example 'gold-plated plant').
- No opportunities for risk-sharing if the project outcome does not match up to expectations.

Advantages of working in-house include better ability to retain and develop key technologies and key staff, and the ability to exert greater control over the details of project execution, without the limitations which may be imposed by a contractual arrangement, for instance the need to negotiate changes rather than to impose them.

There may be a perception that in-house work is either more or less costly than work which is contracted out. Such perceptions, whatever they are, may not really be warranted because of the fundamental difficulties of comparing engineering cost-effectiveness, a subject to which we will return.

If the client does not have the in-house choice due to lack of own resources, or makes the alternative choice anyway, there are still several options. Whichever option he chooses, he will have to retain some in-house (that is, directly employed) resources to manage the contracting-out of the work, and eventually accept the completed project.

If the client wishes to promote competition by making the entire project – the provision of a 'turnkey' plant – the subject of a lump-sum bidding process, he has the task of defining the project work in a suitable way for competitive bidding. For plant of a standardized nature, for example the more standard water treatment, oil refinery, food, or chemical processes, this is a relatively straightforward procedure, and often the preferred option. It is no easy task for a complex non-standard plant – much engineering is needed to develop the conceptual designs and specifications required to define the project work, and there is the inherent problem that in the process of definition, opportunities for innovation and other improvements may be lost.

Some plant owners try to overcome this problem by specifying in detail the plant and services required for purposes of comparative bidding, and allowing bidders who submit compliant offers to submit alternative offers which will not however be the basis for bid comparison. This is indeed either an intellectually shabby concept – why should the bidder submit the products of his ingenuity and competitive technical advantage, if it is not taken into competitive account? – or a commercial charade, if the client does in fact take the more attractive alternative into account, or as sometimes happens invite the lower-priced bidders to re-bid on the basis of attractive technical alternatives submitted by others. These possibilities are easily perceived by the bidders, who can be expected to respond by unimaginative bidding, by various subterfuges (including deliberately ambiguous bids or collusion with their competitors), or by not bidding at all.

Alternatively, the client may foster competition by restricting the plant and services specification to the minimum necessary to define plant performance. This introduces into the bid adjudication process the task of determining how multitudes of plant features will affect the plant's long-term performance, and in particular its reliability and maintenance costs. There is also the risk that this task of evaluation may not be carried out successfully, with the consequence of long-term costs exceeding the short-term savings, or even a plant whose performance or maintainability is unacceptable. Many clients therefore consider this to be an unattractive route, except for simple and relatively standard plant.

The client may try to overcome the problem of long-term performance by purchasing plant on an own-and-operate basis, in which the contractor is responsible for both building and operating the plant, and is paid according to plant output. Lack of competition to provide such a facility (unattractive to contractors because it ties up capital in non-core business) may offset any foreseen gains and, besides, the client may be breeding his own competition to his core business. There are hybrids to these practices in the form of plants purchased with long-term guarantees, but these carry complications of their own, in particular to the management of maintenance and plant improvement without voiding the guarantees.

Another problem with the concept of getting competitive bids for complex non-standard process plants is that considerable time and expense is involved in preparing a bid. The plant must be designed, and the components and their on-site construction priced, in order to submit a bid. To do this properly and accurately may cost each bidder as much as 2 per cent of the plant value, in a business where profits for a successful project may be as low as 5 per cent. Or the bidders may reduce their tendering costs by abbreviating the work and adding a contingency to their price. In the long term, the client can expect to pay these costs and, besides, the whole bidding and bid evaluation process may represent an unacceptable delay to his eventual plant commissioning date.

Why are these problems peculiar to the process plant construction industry, more so than for other types of major investment? The answers lie in:

- the diversity of plant design, including the utilization of many different equipment items of proprietary design;
- the lack of standardization which often is a feature of the optimal design for a given application;
- the complexity of plant operation and maintenance with variable

feedstock and product slate, which often makes it difficult to prove or disprove performance standards;
- the impact of geographical and topographical location, especially for metallurgical plant; and
- the tight project schedule customarily required to suit a changing product market.

Considering some of the issues raised, there are several other contracting structures and hybrids available to the client by which he may seek to balance the conflicting needs of getting competitive bids, getting the best technical solution, and getting his plant built quickly. Essentially, the activities that go into plant design, the purchase of plant equipment and component parts, the construction, and the project management are subject to fragmentation and may be paid for in different ways.

The basic process design – essentially the process technology package as previously discussed, plus follow-through activities to assist and confirm its implementation – may be carried out by the client, or may be obtained from a third party.

If the process technology is not provided by the client, it may be provided together with the detailed engineering and management services, or as a separate entity. We will generally assume only the latter case, as the consequences of combination are fairly obvious. The quality of the process technology is crucial: the whole plant is based on it, and any fundamental error may mean that the whole plant is scrap or uncompetitive in operation. The client therefore needs the most committed participation, and stringent guarantees, that he can get from the technology supplier, and still has to exercise utmost care in ensuring that the technology is based on adequate relevant experience and/or pilot plant tests. The technology supplier's guarantee obligations seldom cover more than a fraction of the plant cost.

By its nature, then, a process technology package usually has to be negotiated as a specific item with appropriate guarantees, and possibly a form of royalty, which will give the supplier some performance incentive.

Engineering, procurement, and construction management may be structured as a separate contract ('EPCM services' or 'management contract'). The contractor's task is the engineering and management of the project and in the process competitively procuring the direct field cost (DFC) elements, that is, all the physical plant and its construction, on behalf of the client or on a reimbursable basis. The 80-or-so per cent of the plant costs composing the DFC are thus subject to competitive purchase, which can be under client surveillance and direction, and only the EPCM services costs need further attention as to how they will be competitively obtained.

These services can be (and are) structured and paid for in a number of ways:

- by the 'manhour' (still usually described as such at the time of writing, but surely under threat of conversion to 'personhour'?);
- by lump sum;
- by percentage of the constructed value of the plant;
- with bonus for good performance, and penalty for poor performance.

There are many variations; we will consider only the major factors which influence the choice.

The performance of the EPCM services impacts directly on the DFC, which is much greater. It is not in the client's interests to save a relatively small amount on the EPCM cost, and lose a larger amount on DFC and plant performance as a result. This is the principal argument for contracting on a manhour basis: there is no disincentive to the EPCM contractor to spend more manhours when justified. Correspondingly there is no incentive for efficiency, and while various ingenious formulae have been drawn up to give such an incentive, it is difficult to overcome the fundamental antithesis between restricting EPCM costs on the one hand, and ensuring adequate work to get best value out of direct field costs on the other. Another problem is that it is very difficult for the client to compare contractors' bids. In an hour's worth of 'manhour', how much is included of dedication, of efficient methodology, of truly relevant experience, of real value?

An EPCM contract on a lump sum basis does not solve the problem of antithesis – there is now every incentive for the contractor to minimize the EPCM input, but none to promote better EPCM performance. This goes even further in the contract, where the EPCM contractor is reimbursed by a percentage of DFC; there is now a positive incentive to raise the DFC. If this can be done in ways which reduce the EPCM input, for instance by purchasing large packages or choosing more expensive suppliers and contractors who require less surveillance or expediting, the EPCM contractor scores twice.

One of the most common systems to seek a balance between the conflicting needs is to split the EPCM work into two stages: the feasibility study and the project. For the study work, the contractor is primarily motivated by the need to secure the project work, which will follow the study. Following the execution of the study, the project work should be sufficiently defined to permit its execution on a realistically motivated basis. There is now a definition of the plant which can be used as a basis for a turnkey contract, and an established project budget which can be used to set up a realistic bonus/penalty incentive for an EPCM contract. This system also has its weaknesses from the client's viewpoint. Other than the lure of the eventual project work, there is no inherent incentive to the study contractor to set the most challenging project budgets and schedules. The contractor who does the study work has a considerable advantage over any potential competitors for the project work. He has gained familiarity with the project, and has participated in the compromises and decisions made in terms of the plant design, pricing, and construction schedule; competitors for the project work, if successful, may be expected to challenge the work done and invalidate the commitments.

It should be no surprise that many engineering contractors are prepared to do study work at bargain basement prices! Clients therefore have to exercise great care that the study is conducted competently, that project cost estimates do not include excessive contingencies or leeway for scope reduction, and above all that the best quality of conceptual design and innovation is obtained. This is not consistent with getting the work done cheaply.

7.2 The engineering contractor

We now swap over from the client's viewpoint to the engineering contractor's. The contractor's first priority, in the long run, must be to get new business. Marketing is obviously pivotal to any business, but the challenge is accentuated when dealing with project work because of lack of continuity – projects by definition come to an end. The larger the projects, the greater the problem of dealing with uneven workload without breaking up teams and developed relationships, and losing the associated benefits of efficient teamwork and predictable performance.

One of the fundamental needs of marketing is to understand and define the product before developing appropriate strategies. For simple and standardized plant, definition of the engineering contractor's product is relatively straightforward, directly corresponding to the client's ability to specify or agree on exactly what he wants. Such plant is usually purchased as a lump-sum turnkey package, built according to specifications and performance requirements based on existing tried-and-tested models. We will not dwell on this end of the market, save to mention that the considerations applicable to more complex plant generally exist at lower intensity: the concept of a standard plant is remarkably elusive in practice!

For more complex and non-standard plant, which for reasons outlined in the previous section is not purchased on a lump-sum basis, the engineering contractor's product is the provision of services; his problem is to define the services in a way in which he can demonstrate their relative value. The principal challenge in defining the comparative value of the product arises because there is no exact reference plant and project. Even when several rather similar plants exist, the project circumstances generally differ greatly, in ways which mitigate against design standardization and which present entirely different challenges to the contractor. Differentiating factors which affect the magnitude of the challenge facing the project team include:

- how well and how extensively the pre-project work was performed;
- project-specific technical considerations, such as country-specific standards and regulations, and local choices of construction materials and methods;
- site access and logistics;
- availability and quality of construction workers, and possible restrictions on expatriate workers;
- environmental limitations;
- the need to use different sources of process equipment; and
- fundamental differences in the attitudes of different clients – some may be impossible to work with, some a pleasure.

There are also challenges that may be presented by the project budget and schedule which affect every project task and make the difference between an easy project and a near-impossible one.

When no real reference plants exist, because new technology is being applied, the engineering contractor is usually faced with another differentiating challenge: the need to accommodate a high number of process changes during the plant design and construction. Inevitably, the technology will continue to develop, and changes will be required as the impact of detail design decisions becomes apparent to the process engineers.

The major differences between the challenges faced by contractors on different projects make it difficult to compare performance in absolute terms – by reference only to the end technical quality, cost, and completion schedule of the plant. Couple this to the inherent problems of marketing something which does not yet exist, a plant which will be built at some time in the future, and the associated (intangible) engineering and construction services, and you have the inevitable consequence of marketing by image rather than by reality. The bidders are drawn into a process of competitive exaggeration,[1] noise rather than logical argument, pretence and charade rather than truth and simplicity.

These aspects can be severely compounded by features of a major investment scenario. The magnitude of the proposed project budget or the length of the completion schedule (or in the worst case, both) may have a critical bearing on the project's viability, on whether it goes ahead or not, and on the futures of the people and organizations involved. There may be intense pressure among all interested parties to accept over-optimistic estimates and commitments. The same circumstances may apply to the adoption of relatively unproven process technology, or construction at a problematic site in a politically unstable region, or any other risks. The client may be looking, consciously or not, for a contractor who, consciously or not, overlooks the difficulties, the risks, and the budget and schedule restraints. The contractors bidding for the work may well be aware of all of this, but may feel that their organization's future depends on presenting to the client the competitive exaggeration that will win the job. This attitude will be even more likely if the contract entered into has potential loopholes to mitigate any penalties for not meeting improbable targets, or if the contractor is paid for his costs irrespective of the project outcome.

The problems outlined above are generally well understood by major clients and contractors. Many clients are content to accept that competitive procurement is not the best method of selecting an engineering contractor, and seek to develop instead a relationship of trust, similar to the basis of a consultancy agreement. Such relationships are, however, subject to abuse and suspicion, and have most of the disadvantages listed for in-house work. In the long run, alternative contractors are likely to be sought, and the process of competitive exaggeration is repeated.

7.3 The project engineer

In the following chapter, which takes the viewpoint of economic analysis, we will outline some of the difficulties faced by the project engineer in quantifying the value and efficiency of his contribution to the project.

---

[1] Colloquially described as 'bullshit' in the industrial environment!

Irrespective of the difficulties, they are part of the context in which he must live and be able to demonstrate his competitive value. The first problem of so doing is to establish the goals in relation to which his performance will be measured.

Given the occasional inevitability of unrealistically marketed projects, the project engineer is likely to face, sooner or later in his career, the challenge of unreasonable goals (following, if you like, the process described above). Maybe he and his team can rise to the challenge with a superhuman performance, but even if this is the case, how demotivating to have such a performance graded as standard, as just meeting the goals!

The key to survival, and the retention of sanity, in these circumstances lies in being able to reduce the work definition, the technical issues, the budget, and the schedule into a breakdown and plan whose logic is incontestable. This is not meant to imply, for instance in the issue of engineering costs, that there is a magic figure which is the 'correct' cost of engineering, arrived at as a universal truth. It means that the underlying issues can be exposed and the potential implications of either more or less engineering input can be demonstrated. Ultimately, there will always be the possibility that a given task can be done more or less efficiently, but it is only by breakdown to a low level of task, for example the document level, that true comparison can be made. Otherwise, there is always the danger of simply eliminating inputs and thereby creating adverse consequences to another part of the project, for instance the DFCs or plant maintainability. And if it arises that the 'goals', the project budget and schedule, have been arrived at without any such low-level breakdown, then it can be surmised that the commitment is purely a gamble.

We will discuss three basic forms of contract under which the engineer may work, and some of the consequences in terms of how work should be conducted and its effectiveness judged. The first is the lump-sum turnkey job, in which the contractor has total responsibility to build a plant that complies with a specification, within an agreed time-frame. This is in many ways the least frustrating way to work, because the contractor is free to choose his approach to the work, and how to structure and optimize his inputs to arrive at the most economic end result, consistent with technical acceptability and completion schedule.

An essential challenge in controlling the work is to maintain absolute clarity about the scope of work and the plant specification. It is not uncommon for clients to want the best of both worlds, to want to control the engineering of the plant in accordance with their ongoing preferences, but to maintain a fixed price and schedule. This is obviously a nonsense: a fixed price for an unspecified product. The most common device for clients to attain this end is to specify that project work shall be 'subject to approval', either without further qualification, or more usually with reference to cover-all phrases such as 'good engineering practice' or 'state-of-the-art design'.

Clearly any project engineer who wishes to avoid a sticky end must insist on clarification of such open-ended terms, before advancing too far into the work. Failure to do so is no different in principle from advancing into detail engineering without freezing the conceptual design. There is no knowing the outcome. Despite the possible negative perception by the client, it is essential to agree a fair definition of the basis for approval, and this is always easier to achieve prior to the arrival of a contentious issue. A fair basis is not baskets of specifications such as 'all relevant national and international standards', but one in which approval is based on stated parameters and specifications without the imposition of arbitrary preference, which, if required by the client, should be the subject of provisional pricing or change orders.

The second contracting mode is the hourly paid EPCM contract. Here the potential for conflict between contractor and client is minimized, and the contractor's job is simply to perform professionally. There is generally correspondingly little incentive for the contractor to perform above expectations, and there may be strong pressure to minimize the input costs per manhour by minimizing staff costs – not a happy position for the manager charged with maintaining quality. There is little scope for manoeuvre in this situation other than the negotiation of a bonus according to comprehensive performance targets, which effectively hybridizes the contract with the next type to be considered.

Finally, there is the lump-sum EPCM contract, arguably the most challenging in that there is a fundamental adversity between client and contractor. Once the lump sum has been agreed, there is no financial brake on the client's aspiration to reduce DFC and make ongoing improvements to plant design by demanding additional work by the EPCM contractor. It is just as important to maintain a rigorous control of workscope as for turnkey work, but it is more difficult because the product is less tangible. To control the engineering workscope, both the activities (and, by inference, the number and type of documents) and the definition of work acceptability have to be rigorously controlled. This involves the strict application of the precepts outlined for planning engineering work, in Chapter 4, and in particular the following.

- Develop and obtain client approval for comprehensive design criteria and design approval checklists before doing any other work. Try to ensure that design criteria and checklists contain only clear requirements and no debatable issues, that is, no words such as 'best practice', 'optimum', or 'vibration-free', which are likely to lead to controversy through lack of quantifiability.
- Work by increments: do not commence with detail designs until conceptual designs and layouts are approved.
- Strive to maintain a regime under which any proposed design change is regarded as a change to the scope of the contract, unless the change is necessary for plant performance as specified, or needed to satisfy safety criteria.

7.4 Conclusion

Some of the views, concerns, and practices which have been discussed may be regarded as confrontational and unnecessarily pessimistic of human nature. In many situations this may be the case: engineering work may be conducted in a harmonious and professional manner without any client/contractor conflict; relationships of trust may be developed without any abuse or loss of competitive performance. However, the project engineer must be aware of the full range of working relationships and their consequences, if only to reinforce his determination to maintain the *status quo* of an existing relationship.

On the other hand, it is entirely possible for an engineer to find himself in a situation which cannot be managed in terms of the logical methodology discussed. For instance, to be locked into a contract with a defined price, defined project schedule, undefined scope of work, and an unreasonable client. The only response to this situation is firstly to analyse and expose the basic unmanageability of the task, and then to negotiate, preferably before doing too much work.

Chapter 8

The Economic Environment

In this chapter, we will further address and quantify some of the 'project environment' issues raised in preceding chapters. They essentially relate to two aspects of project work, firstly plant capital and operating costs, and secondly the costs of project engineering and management work.

8.1 Plant profitability

One of the main objectives of a plant feasibility study is to assess the commercial viability of the proposed plant. The essential parameters of this calculation are the capital cost and cost of debt, the operating costs, the revenue on operation, and the plant life. It is also necessary to evaluate the risk that the values of these parameters may change.

The parameters of initial cost, operating cost/revenue, and plant life can be brought together in many ways. One method is to determine an annual amortization cost corresponding to the initial plant capital cost such that the capital and interest[1] will be repaid over the operating life, considering also any decommissioning costs and final plant disposal value. When this annualized cost is added to the direct[2] operating costs, the total may be subtracted from the

---

[1] The interest can be calculated in different ways which are equally valid, provided that the users of the calculated values understand the significance of the different calculation methods. One way is to use the interest corresponding to the cost of debt capital to the client organization. Another way is to use the rate of return expected for equity shareholders; another is to use the *opportunity* rate, the rate of return which may be expected from alternative investment possibilities.

[2] 'Direct' in this context means all costs other than those arising from the capital cost and associated interest.

operating revenue to yield the net annual profit expected from the plant venture.

Another method of analysis is to compute the net present value (NPV) of the future plant operating margins (that is, product revenue less direct operating cost), by discounting future earnings at an appropriate interest rate,[3] and comparing the NPV with the estimated capital cost. It is a simple mathematical exercise to demonstrate that the two approaches are merely different presentations of the same information. In the following, we will use the NPV approach, because it is easier to utilize when making decisions on different plant features which have capital/operating cost implications.

8.2 Lifecycle considerations and 'trade-off' studies

In the overview on project management, we referred to the concept of a plant lifecycle, of which the project is a part. In the following, we will discuss some of the 'lifecycle' aspects which have to be considered when conducting a study and building a plant.

Project design criteria include the need to consider plant operability, maintainability, corrosion protection, and various other features which improve plant operation, operational safety, availability, and life. Typically, these features have a minimum level of implementation that is mandatory, below which plant operation is unworkable or unsafe or even illegal. On top of this minimum, there is a level of 'reasonably expected' performance – for example, the life of components before failure – for which the engineer may be held to be negligent if the performance level is not achieved, through no fault of the operators. These 'reasonable expectations' may be governed by the client's specifications, industry practice, the reputation of the engineering organization, or common-law interpretation, and should be understood and considered (and qualified if necessary) when drawing up design criteria.

Further operational performance enhancements are generally a matter of balancing the capital cost of the enhancement against future operational cost savings.[4] Of course, these enhancements (or the plant features which secure them) may be specified by the client, in which case

[3] The same considerations about the rate apply.

[4] Or increased revenue arising from more, or improved, product. If revenue is regarded as a negative cost, the treatment is identical.

there is no need for further discussion. But typically at the study stage, it will be the engineers' responsibility to determine the correct balance. Theoretically, this may be established by calculating the net present value of a future-cost-saving option, allowing for the fact that the future savings have to be discounted by a rate corresponding to the effective interest rate, which is the cost to the investor of financing the option. For large sums, arriving at the effective rate and making the decision is often no simple matter; it may depend on the availability of venture capital and, indeed, the additional finance may not be available whatever the return. For smaller sums, it is usually possible to obtain a suitable rate for the purposes of decision-making, and it is a simple matter to compute the net present value according to the discounted returns over the life of the plant.

For everyday decision-making, it is convenient to express the result in terms of a time pay-back period by which to ratio annual cost-savings to arrive at the discounted total of future savings. Having evaluated this figure once for the project, it becomes a criterion which can quickly be applied to each application.

As an example (there are variations according to the financial presentation required), if

P = net present value of total future savings
S = net cost-saving per year
n = plant life (years)
$i\,\%$ = annual 'effective interest rate' (or discount rate)

Then if we define d by

$$d = \frac{100 - i}{100}$$

$$P = S(1 + d + d^2 + d^3 + \cdots + d^{n-1})$$

Multiplying by d and subtracting to sum the progression

$$Pd = S(d + d^2 + \cdots + d^{n-1} + d^n)$$

$$P - Pd = S(1 - d^n)$$

Therefore

$$P = S\frac{1 - d^n}{1 - d}$$

$$\frac{P}{S} = \frac{1-d^n}{1-d} \text{ years}$$

For instance, if the annual discount rate is 20 per cent ($d = 0.8$) and n is 15 years

$$\frac{P}{S} = \frac{1-0.8^{15}}{1-0.8} = 4.82 \text{ years}$$

In the above, we have assumed that the benefits of 'the feature' are available over the life of the plant, which implies that any additional maintenance or periodic replacement costs to keep the feature effective have been deducted in calculating the annual cost savings. Alternatively, of course, if the proposed feature has a life of say 4 years, then one can simply put n equal to 4 to evaluate the proposal. It is also a simple matter to revise the algebra and series summation above to reflect a different incidence of cash flow, for example reflecting commencement of repayment in the second year after the capital outlay

$$P = S\frac{d-d^n}{1-d}$$

In the general case, if cash flow arising from the investment expenditure commences a years after paying for it, (a may be zero or one as above, and need not be an integer)

$$\frac{P}{S} = \frac{d^a - d^n}{1-d}$$

In practice, the accountants may further complicate the exercise to justify their existence, but the same principles apply. (This statement is of course unfair to the accountants! They are obliged to make the analysis follow the dictates of various wretched taxation authorities and bankers, on which there is further discussion below.)

However, the financial analysis is simpler than the actual application. The problem is to establish the annual savings accurately and meaningfully, and plant operators are, out of experience, inclined to be sceptical about claimed savings – in effect, to apply a further discount to them.

For instance, the engineer may be confronted with a choice of two pumps, one with better efficiency than the other, leading to an apparent power saving of 10 kW at 8c/kWh, the benefits of which saving commence immediately after the capital outlay.

It is superficially easy to say:

Plant utilization factor = 0.95
Value of power saved per annum = 0.95 × 365 × 24 × $0.80/h = $6658
Pay-off ratio (as per example above) = 4.82 years
Maximum justified capital cost increase = $6658 × 4.82 = $32 090
(which may well exceed the cost of both pumps!)

In practice, there are reasons to doubt whether this is realistic. The pump vendor may be over-optimistic in his efficiency statement, knowing that the efficiency claim is unlikely to be closely verified. Accurate performance testing in the shop is an expensive addition, and there is always a testing 'tolerance'; it is even more expensive and inconvenient to carry out individual pump performance testing when the plant is operating. The actual service duty (head, flowrate, specific gravity, etc.) of the pump is likely to differ from the design case, and the performance will vary accordingly. (The specified performance invariably includes margins which are not completely utilized; the plant operators may decide that the plant operation is more stable at different conditions, etc.) Wear, corrosion, and maintenance practices may have an even greater effect on power efficiency, for example by increasing the clearance between the impeller and casing wear rings. Such considerations make it more effective to evaluate possible cost-reduction modifications once the plant is fully operational, and its performance has been evaluated and optimized.

This certainly does not mean that the process outlined above is unnecessary at the initial design stage; the 'other side of the coin' is that it is more economical to make changes in the initial construction than to retrofit. The intention is to caution that it may be advisable to take a jaundiced view of potential savings, for example by applying an increased discount rate, and to be circumspect in evaluating cost savings due to better performance unless the savings are rigorously demonstrated by testing. In any event, the approach to be taken should be clearly agreed with the plant owner at the stage of design criteria development.

Evaluation of future maintenance and replacement costs of equipment tends to be even more debatable than evaluating increased efficiency. Once again, equipment suppliers may be expected to make exaggerated claims for their product in a competitive market, but here the actual performance is even more difficult to confirm. The quality of plant operation and maintenance and the possibility of unrecorded or undetected abuse, as well as deviations between specified and actual operation, make it rather difficult to apportion responsibility for equipment maintenance costs and indeed failure, over the longer term. It is too easy for the equipment supplier to blame the operational environment for poor performance, and it is no wonder that equipment is usually purchased on a 1-year guarantee period, even though the life expectancy may be 15 years. Consequently, it is more normal to specify equipment features which should result in increased reliability and longer life, rather than leave the emergence of such features to the competitive purchasing process. In fact, the nature of a process plant is such that potential causes of accelerated failure and high maintenance simply cannot be tolerated. It is customary in the equipment selection process simply to eliminate options which are less desirable on these grounds rather than try to evaluate the cost–benefit trade-off.

Not that the maintenance costs can simply be overlooked in equipment selection. In particular, the cost of spare parts, and as far as practicable the equipment supplier's policy and reputation in regard to spares and maintenance costs, need to be considered. Several manufacturers aim to make all their profit on spares and none on the original equipment.

For study purposes, it is of course necessary to estimate plant maintenance and periodic component replacement costs for the overall plant economic evaluation. This is usually made on a basis of experience of similar plant and environments, for example 4 per cent of total plant capital cost per year. Estimates made on the basis of detailed build-up are inclined to be too optimistic and not worth the trouble of the exercise. This may not be the case for single pieces of large equipment, or plant which is centred on relatively few major equipment packages, as for power generation.

Decommissioning costs may also be a factor to be considered in economic evaluation, but deferred cash flow discounting usually greatly reduces the significance. An exception may be plants which create lasting pollution, but such plants are increasingly unlikely to be licensed for operation in the first place.

8.3 The real world of costs and values

We touched above on the impact of fiscal authorities and capital sources on the economic trade-off which determines the optimum choice of many plant design features. Their impact is such that they are worthy of further elaboration.

Taxation rules that affect plant finance are many. The most important factor is usually the treatment of depreciation, which is frequently used as an investment incentive by allowing the capital cost to be written off against operating costs over an unrealistically short period, for example 3 years (notwithstanding that maintenance policies are usually to maintain or enhance plant value in the first years of operation). The effect of this is to make capital cheaper, by an amount which depends on the discounted value of the tax saving.[5] On the other hand, there may be a tax 'holiday' over the period of initial operation, effectively nullifying the benefits of depreciation. The taxation treatment of items of operating expense and revenue may likewise have an effect on the optimum plant configuration and design.

Low-interest finance packages, such as export incentives, may also have a major effect on the cost of capital, and influence the design more directly by restricting the sources of supply. Thus in the real world, it is usually necessary to work out each year's cash flow in detail, and to apply actual taxation and item-specific capital interest, to reach the correct trade-off for individual item design optimization. Failure to ascertain the real impact of the fiscal and investment environment can lead to quite uncompetitive designs.

There are some factors that affect the capital/operating cost dynamics which are best left to the plant owner. One is the assessment of uninsurable risk, related to the possible effects of changes of cost inflation and changes of the market for the plant's product. The second is that in the wider field of the client's operations, especially for global companies, it may be expeditious to incur apparently sub-optimal costs and profits at isolated parts of the overall operation, in order to centralize profits in one place rather than another, and thereby make significant tax gains.

However, if the correct attention and consultation is given at the outset of the project or study, it is usually possible to arrive at a simple set of rules, expressed in pay-off time, that makes financial analysis relatively simple. (The reservations expressed above on the credibility of certain technical aspects of the analysis remain.)

8.4 Project engineering and management work

For the economic analysis of engineering and management work, it is first necessary to recall the observations made in Chapter 5, The

[5] There is occasionally a sting in the tail here; if the plant performs subnormally in its initial years, there may be no profits on which to reduce the tax.

Project's Industrial Environment. In particular, that the work of the project engineering and management team (costed as indirect field costs or IFCs) may be segregated into a core of essential activities and a variable amount of work which can only be justified if the benefits of the work exceed the costs.

Much engineering work is an iterative process, in which designs are conceptualized, developed and presented in a comprehensible format such as a drawing or model, reviewed, and then re-worked until approved. There is no inherent limit to the number of iterations of review and re-work. As projects become larger and more complex, with the participation of a greater number of disciplines carrying out interrelated activities, the need for correspondingly more review and consequential re-work becomes apparent. Failure to do so results in construction clashes, interface errors, and equipment accessibility problems. The costs of review may be reflected in additional manhours, or in additional activities of physical and computer modelling, process simulation and testing, and also in the cost of schedule delays while additional engineering work is done as a consequence of the review.

By economic consideration, the optimum amount of review and re-work, and hence the optimum amount of engineering, has been performed when a trade-off is reached between the benefit of review and the associated costs. The benefit, for this purpose, is the added value in terms of improved plant design and reduced construction cost with reduced error rectification.

The trade-off analysis can be expressed formally by the standard elementary economic theory of diminishing returns. It is illustrated in Fig 8.1 which shows the curve of direct field cost (DFC) against engineering cost (EC). 'Diminishing returns' says that the curve has a negative slope which is constantly increasing, asymptotically, to zero slope. The line of engineering cost against engineering cost is obviously a straight line with slope +1; adding the two curves yields the total cost (TC) of the engineered plant. By inspection or by differentiation, there is a minimum total cost when the rate of increase of DFC with respect to engineering cost is equal to –1, in other words, the incremental cost of engineering equals the associated decrease of DFC.

Care needs to be exercised when making historical comparisons of the engineering costs of projects on what was the engineering workscope on the job in question. As we have previously seen, the engineering and management scope can be shifted between the project engineering team, the equipment suppliers, the fabrication and construction contractors,

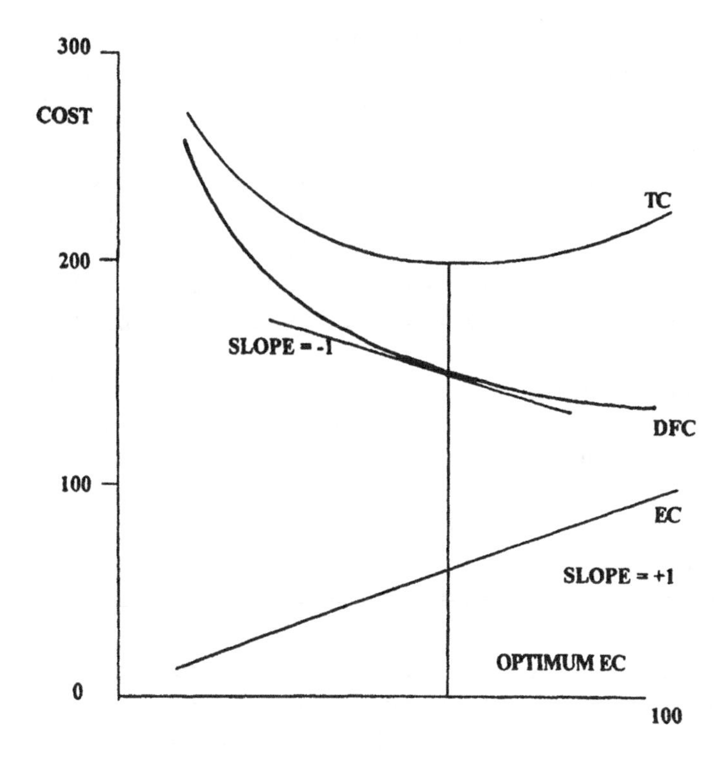

Fig. 8.1 Optimization of engineering inputs (The same logic applies to all management inputs)

outside consultants, and the plant owner's team. More specific examples of this practice include:

- the purchase of integrated packages of equipment rather than individual items;
- omission of drawings, such as piping isometrics or as-builts, which become the responsibility of the fabricator;
- using lump-sum rather than per-item contracts for construction;
- imposing field measurement checking responsibility on the contractor; and
- reducing office checks.

The list above is potentially endless. The same overall engineering work is done, but the work is shifted to the direct field cost report.

Extra engineering work can be created by imposition of exacting design and documentation standards, and reduced by the acceptance of previous project or standard designs. Procurement policies can have a major effect on the engineering work required for an order. The engineering work of construction support, manufacturing follow-up, and commissioning activities can be defined and reported in different ways, all of which will impact on the perceived engineering workscope.

Work can also be shifted between the study which precedes the project and the project itself. The owner/investor can decide as a matter of principle that the design basis produced for the study is frozen at the outset of the project, or that it must be re-worked completely as part of the project work.

It should be noted that the various ways of carrying out the project, and their impact on apparent engineering costs, are not just crafty devices for reducing or increasing engineering costs (although that may be the intention!). The best option may be different on different projects or for different contractors. There is no golden rule as to what is optimal.

Assessing engineering work as a percentage of plant capital cost can be misleading in other ways. If the design capacity of a plant is doubled, without any change to the complexity,[6] there may be little or no change to the quantity of engineering work – the same number of documents have to be produced, with the same number of information items. But the capital cost of the plant can be expected to increase by a ratio of something like $2^{2/3}$ (or 60 per cent). On this basis, the curve of engineering-cost/plant-cost ratio versus plant capacity will feature a continuously decreasing ratio as capacity increases. However, the concept of such a curve is more likely to be misleading in practice, because the process of trade-off (if allowed to take place) will dictate that extra engineering activities are economically justifiable on the larger plant. It is worth spending more time making improvements.

In conclusion, it can be very misleading to set standards or draw statistical comparisons on what should be the 'correct' number of manhours or engineering cost for a given project. There can of course be large differences in performance between different engineering teams. Engineering performance is affected not just by the manhour input, but also by all the usual factors of skill, training, effort, use of advanced systems and software, availability of previous usable designs, and good management. One would expect a good return from better engineering, and arguably this can only be gauged from an expert assessment of how the work is done in relation to the challenges of a particular project, rather than by drawing up statistics on cost percentages.

---

[6] Complexity is usually assessed by the number of mechanical equipment items.

The general observation about increased engineering work providing diminishing returns, and thus leading to an optimum at which the overall project costs are minimized, is in fact general to all project management activities. There is an optimum value for the intensity of each activity at which the marginal cost benefits become less than the cost of the input. As the way in which each cost element responds to management input varies according to the circumstances of each project, and can even change from day to day (for example, as extra project management and engineering input are needed to solve supply problems), there is clearly no unique overall optimum value – 'the right cost for project management for this type of project' – except perhaps in cloud-cuckoo-land.

Third Cycle
Conceptual Development

Chapter 9

Studies and Proposals

9.1 Feasibility studies

In the preceding brief overview of project management, we chose to address the workings of a project as a task with defined goals and a defined and accepted concept for achieving them, with only the most cursory mention of the work necessary to get to that point, referred to as 'the study'. Now we will address the means of getting to that point. There are a few reasons for moving in this apparently bizarre fashion. The ultimate objective of the study is the project, and it is easier to comprehend many of the study activities if we start by understanding the objective. Much (but not all) of the work executed in the study is in fact an abbreviation of the more detailed work carried out at the project stage, when physical commitments are produced rather than paper abstractions; it is difficult to properly describe an abbreviation without first describing the full process. Competent study practitioners have first to become competent project practitioners, although the full learning experience seems to be a number of cycles of both.

We will begin from the point when a management or an investor takes a decision to expend resources to examine the business potential of an idea, in other words, commissions a study. This is at any rate the starting point for the purposes of this book; it would be possible to digress at great length on both technical and business research processes leading up to the production of worthwhile ideas.

The study invariably includes the following.

1. *Defining the idea.*
2. *Developing the process application of the idea.* The work to be done obviously depends on what has gone before. It may be necessary to

commence with laboratory research and bench-scale testing of processes, and with geological exploration and widespread sampling of raw materials. Proving the reserves of acceptable plant feedstock is often a major part of a study, but it is not addressed here because our interest is in the process plant.

In general, it is necessary to obtain sufficient data on the materials to be processed, consider the available process options, carry out any process testwork that may be required to demonstrate workable process routes, and prepare flowsheets based on the applicable processes.

3. *Evaluating the process proposals technically and economically.* The technical evaluation may involve further testwork, and possibly the construction of a pilot plant. It may be possible to observe similar existing plants, where the performance of the process can be demonstrated under industrial conditions. The economic evaluation implies estimating the capital and operating costs of the plant, and for this it is necessary to prepare at least an outline design of the plant. The economic evaluation may also involve product marketing surveys and similar activities beyond our scope.
4. *Reviewing, testing, redesigning, optimizing, and re-evaluating the various possibilities*; selection of the most attractive proposals, and presentation of the corresponding technical and economic evaluation.

Usually before a project goes ahead, investors demand a level of confidence in its technical and economic viability, in keeping with the value of the investment. Before a major project is authorized, it is not unusual to go through years of technology search and negotiation, laboratory and pilot plant tests, plant design and costing exercises, hazard and risk analysis, environmental impact studies, market studies, and financial analysis. There may also be negotiations with statutory and government bodies, trades unions, buyers of the product, and other interested parties.

As our focus is on project engineering, we will direct our attention to those parts of the study with which the project engineer is most concerned, beginning with the evaluation of the process and the process design information. Firstly we need to know that we have a process which works, and can be operated in an industrial environment. For all but the simplest processes there are only two ways to do this: preferably, by ensuring similarity to existing successfully operating plants, or failing that, by operation of a pilot plant. The challenge here lies in the word 'similar'. To know what constitutes similarity, and to be able to recognize any small difference which may have a critical effect, requires an expert knowledge of the process in question. Thus it is necessary to have both reference and/or pilot plants and an expert opinion on their relevance to assume any level of confidence in an application of process technology.

To design a plant, which can be expected to work as well as the existing plants on which the design is based, usually requires an understanding of several special engineering features which may not be readily apparent. Such features may include, say, special materials of construction, small equipment-design changes, and specific maintenance features, which are the product of the process development, often arrived at through expensive trial and error. We need to have the leading participation of the process technologist in the development of the full plant design criteria.

The development of appropriate process flowsheets and basic process engineering is also outside the scope of this book. We will begin at the point of developing the process designs into actual plant designs, which can be assessed technically and costed. In principle, the work that has to be done is the same as the initial work already described for the engineering of projects:

- comprehensive design criteria are prepared;
- process equipment bids are solicited;
- equipment selections are made for purposes of plant design and costing, but without any commitment;
- plant layouts are developed in conjunction with appropriate materials transport system design; and
- structures, civil works, electrical and instrumentation systems, piping, and any other utilities and facilities are designed and costed.

However, unlike project work, studies are usually commenced without a definite plant design concept and construction plan; these are subject to change, and the essential requirement is to find the most suitable concepts and plans which will become the basis for the detailed project designs, budgets, and schedules. Inevitably the plant design is developed in a series of iterations. Initially, relatively broad concepts are explored, and possible innovations are introduced. Some concepts, for example materials handling system, layout, or site location, may be quickly rejected, while others may need more detailed comparison to arrive at the best choice. The phases of initial conceptual development, and of more detailed design and evaluation, are often formally split into two parts: a pre-feasibility study with an accuracy[1] of say ±20 per cent, and a feasibility study with an accuracy of say ±10 per cent. This gives the investor an opportunity to abort the study work or change direction before too much money is spent.

The development of cost-effective designs is discussed in Chapter 11, Value Engineering and Plant Optimization. Cost estimating is clearly at the heart of study work; it is one of the investor's main concerns. In keeping with the need to carry out initial study work comparatively quickly and less accurately, there are a variety of estimating techniques available to estimate plant costs without doing too much design work. Ultimately, however, an accurate estimate must be based on the submission of competitive bids for all items, and the bids must be based on the adequate specification of equipment items, and adequate quantification of sufficiently developed designs of other items.

9.2 Proposals

Studies may be regarded as a type of proposal, in that they relate to the proposed construction of a process plant. However, we will restrict the use of the term 'proposal' to that of an offer, namely an offer to build a specified process plant at a certain price or price basis, a commitment rather than an estimate. 'Study' will be used when an estimate of plant costs, rather than a commitment, is submitted. The essential content of a process plant project engineer's work is the same for both, but there is a fundamental difference in how the end product is used, in that the relationship with the client is different. There is a corresponding difference in the assessment of risk.

9.3 Estimating project costs

Certain cost estimating techniques are relevant to pre-feasibility studies only, and will therefore be addressed first. There are various techniques for making rough or preliminary estimates of plant cost which are particularly useful for pre-feasibility studies. The most elementary of

---

[1] 'Accuracy' is a word which is customarily used in this context, but the understanding of what is meant varies quite widely; in fact the word is often used without any understanding of its implications, such that the value of the quoted accuracy is meaningless. The suggested usage is discussed in Section 9.3.

these is the 'curve price', in which the entire plant cost is interpolated or extrapolated from data on previous projects to construct similar plant, the 'curve' in question being a graph of plant cost against capacity. Due to limitations on similarity, it is seldom possible to assign an accuracy of better than ±30 per cent to such an estimate.

The more developed techniques for preliminary estimates are based on factorization. The basis for factorization is the observation that the costs of component parts of process plants bear similar ratios to each other in different plants.

Consider for instance the breakdown of the plant capital cost into the following (all components include the associated site construction and painting costs):

- civil works
- structural steelwork
- mechanical equipment
- electrical equipment and reticulation
- instrumentation and control gear
- piping
- transport to site
- indirect costs (engineering and management, insurance, etc.).

This breakdown includes all elements of the plant, that is to say that possible other elements of breakdown (such as platework and valves) are included in the above headings (for example mechanical equipment and piping respectively).

Now the largest of these components is invariably the mechanical equipment, and it is also the most fundamental component, being arrived at directly from the process flowsheets and process requirements. The other plant items follow from the mechanical equipment needs. So in the simplest factorization technique, the plant cost is factorized from the mechanical equipment cost, which is generally in the range of 30–45 per cent of the direct field cost, or 25–40 per cent of total cost, including indirects.

There are many possibilities for improving the accuracy by designing and estimating more of the other plant components, and factorizing only the most intractable, say piping. The technique is also refined by developing ratios of plant costs corresponding to each type of mechanical equipment, for example the ratio of the cost of a centrifugal pump to the cost of the associated civils, structurals, piping, electrical, and instrumentation. An appropriate factor is developed for each mechanical equipment type; the plant cost is the sum of the individual mechanical equipment costs, each augmented in its appropriate ratio.

It will be evident that the accuracy of factorization techniques depends on the similarity of the type of plant, and also the consistent grouping of cost elements into the factorized components. A plant which is intensive in bulk solids handling is not comparable to an oil refinery. If the cost database is inconsistent as to whether, say, thermal insulation costs are included under piping on one project and under mechanical equipment on another, accuracy will be reduced. When there is a reasonably consistent database of reasonably similar plant, factorization of the mechanical equipment total may yield an estimate for which an accuracy of ±25 per cent can be claimed. This could be narrowed to say ±20 per cent if the more advanced factorization techniques are used, but hardly better unless there is extreme confidence in a recent database of very similar plant.

For most clients and circumstances, the level of confidence required, before the project is authorized, dictates that the plant is designed and costed in detail. Therefore factorization-estimating methods are usually restricted to pre-feasibility work or comparison of plant alternatives.

'Costed' means that costs are estimated in a way that should correspond to the agreements that will eventually be struck with suppliers and contractors when the project goes ahead. The most obvious way to arrive at such prices is to solicit bids based on appropriate specifications, and to review the bids technically and commercially as carefully as for an actual project.

The knowledge that costs are based on actual bid prices is one of the principal sources of confidence in the estimate. However, there are limitations to this practice. Suppliers may object to being used in this fashion, when no immediate or perhaps any potential business can arise from their effort, so they may decline to bid or submit uncompetitive offers. The work of specification, enquiry preparation, and bid analysis may overload the study/proposal budget and time schedule. Cost-estimation techniques, databases, and experienced judgement are therefore essential supplements to direct market information.

Apart from the need for job estimates, individual engineering disciplines have to develop and maintain current databases and methods for costing plant design alternatives in order to select the best designs. This process has to be carried out so frequently and rapidly that it is not practicable to be wholly dependent on solicitation of market prices whenever data are required. In fact, an engineering organization where the individual engineers have little relevant cost data or cost-estimation techniques is invariably not a cost-conscious organization, but rather an uncompetitive monolith.

It is clear therefore that a good knowledge of cost-estimating techniques, and maintenance of a relevant database, are an essential part of the project engineer's armoury. There is no lack of publications on the subject; for instance, the guide produced by the Institution of Chemical Engineers and the Association of Cost Engineers is a useful introduction.[2] However, at the overall project or organizational level, there is no substitute for the employment of experienced professionals for this function.

9.4 Risk

The engineering and estimating work of a study or proposal is not complete until its accuracy has been established. For pre-feasibility-type work, where the level of commitment is relatively low – at most leading to a decision to finance a full feasibility study – it is usually considered acceptable to presume an accuracy based on experience; that is, based on previous validated experience that enough design and costing work has been done to justify the accuracy quoted, which is unlikely to be better than ±20 per cent. It is also expected that any significant uncertainties and hazards will be summarised in the pre-feasibility study report, for detailed attention during the feasibility study.

However, when a firm commitment to build a plant is under consideration, detailed assessment is normally required of all cost elements and of everything that can go wrong and affect the viability of the proposed commitment. The process is described as risk analysis. Many treatises and methodologies are available, as befits the gravity of a subject which can seriously affect the fortunes of major enterprises. We will address the fundamental aspects.

A risk is defined as a possibility that a project outcome may differ from the planned outcome.[3] Risk can be quantified as a probability. The

[2] Institution of Chemical Engineers and Association of Cost Engineers (UK): *Guide to Capital Cost Estimating*, 2001.

[3] There are other definitions, for example 'the likelihood that an accident or damage will occur', which is more appropriate in a safety context. In the commercial context, there is also risk of a better outcome (lower price, better performance), which must be considered in quantitative decision-making.

principal project outcome is financial, and the consequences of many other unsatisfactory outcomes – late completion, unreliable operation, unsatisfactory process performance – can be expressed in financial terms, either as cost of rectification or reduction of plant profitability. Some possible outcomes may be unacceptable in their consequences, and it will be necessary to take action to avoid, manage, and mitigate the risks.

The first step in the process is risk identification, which is accomplished by review of the proposal/study by suitably experienced experts using well-considered checklists. These must embrace all facets which may affect the project outcome, such as plant feedstock quality, process reliability, mechanical reliability, potential foundation problems, operational hazards, environmental impact, equipment costs, construction costing and labour, statutory requirements, contractual and legal problems peculiar to the country of construction, eventual decommissioning costs, and so on. Risks due to operational hazard are generally treated separately from commercial risk, and will be discussed in Chapter 12, but they are obviously an essential part of the overall process.

As risks are identified, the potential impact is assessed, and if considered significant, risk management action is decided. This could be the commissioning of additional testwork, hiring of appropriate experts, sharing of risks by contractual arrangement, introduction of more conservative designs, taking out of insurance – there are usually many possibilities. Most risk management actions simply become another project activity, with an additional cost component in the proposed project budget.

Quite apart from the significant identified risks, there is always the possibility of unidentified risk, and there is always a risk that individual cost components within the project cost estimate may eventually prove to be incorrect. Design development, inflation, changes in the suppliers' marketplace, and the difficulty in making firm supply and sub-contract commitments at the time of estimating mean that most items have to be repriced at the project stage. This is not a one-way process; often there is more intense competition, and lower prices are available. These pricing uncertainties, and the variation of possible cost impacts arising out of the risk assessment, have to be addressed by experienced judgement (possibly aided by statistical analysis) to arrive at an appropriate contingency; this is discussed in Section 9 below.

At the conclusion of the process there will be a report on risks identified, potential impact, management action considered to be appropriate, and an evaluation of whether the acceptance of the managed risk is considered to be 'reasonable'. This is about as far as the risk analysis and management process can go, unless there is an overall statutory or corporate dictum that certain risks are not considered to be tolerable. Ultimately, a risk-taking executive, body, or board has to make an informed decision on whether to accept the risk, in the light of the rewards for adopting it or the cost of the opportunities lost.

As a final note, project engineers need to bear in mind that a legal risk is inherent in any of their work. The more obvious obligations to exercise reasonable care and diligence in producing designs which are safe to build and operate probably require no emphasis. However, in making a recommendation to an investor (even indirectly), there is an obligation to exercise care and diligence, which can attract very substantial civil liability. Depending on the country of investment, and possibly any other countries in which investors may be affected, the obligations may be quite unreasonable in relationship to the engineer's fee and scope of work. Legal advice, qualification of responsibility, limitation of liability, and insurance should be considered before making any commitment or submitting a report.

9.5 Accuracy

The 'accuracy' of an estimate is a widely misused and misunderstood term. Mathematically, accuracy can be expressed by the limits of a range of values within which the correct figure lies. If the only possible solutions to a problem lie within the range 6–8, then the answer may be expressed as 7 ± 1. However, life is not so simple for the plant cost-estimator. His estimate is made up of a large number of cost elements, none of which is certain, even as to the *absolute* limit of the range of possible cost. There is no certainty in his life, only probability and confidence. Certainty only exists when the project is over!

In fact, the estimate can best be presented as a probability curve, a plot of estimated cost (x-axis) against the probability that the actual cost will fall within a defined band (say ±2 per cent) of the estimated cost (y-axis), which is shown in Fig. 9.1. The curve will clearly feature a maximum value at the most probable cost, or the cost considered to be most probable, and slope downwards on each side of the maximum. That is about all that is known about the shape of the curve. Many models (normal, skewed, Poisson, geometric distribution, etc.) are assumed and used in practice.

Because it is a probability curve, sloping away on each side to diminishing but never-zero values, it is only possible to assign limits

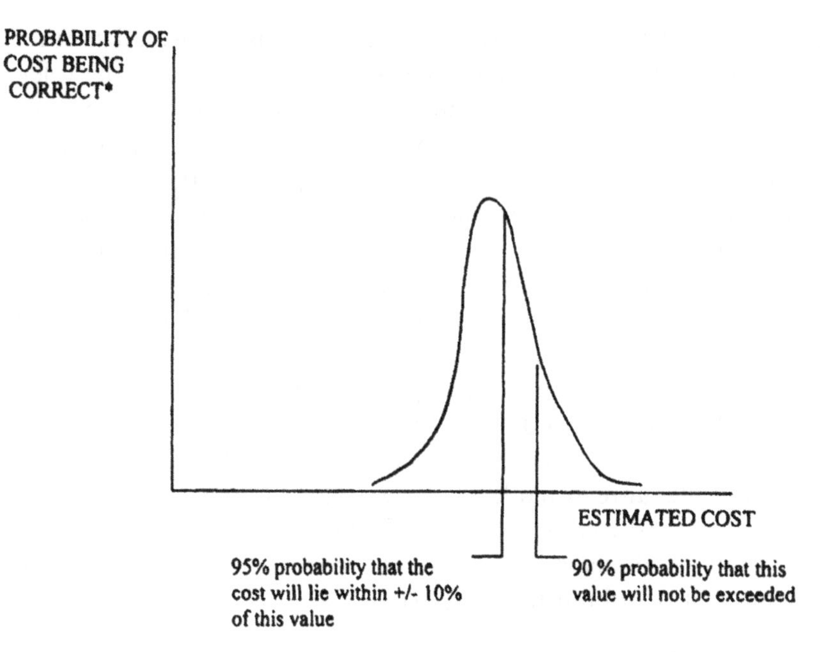

Fig. 9.1 Plant cost probability

of accuracy in association with defined probability or confidence levels. Thus to claim a plant cost estimate is accurate to ±10 per cent is meaningless in itself, unless a confidence level is assigned to the limits. It is, for instance, meaningful to claim that the estimate is accurate to ±10 per cent *with 90 per cent confidence that the estimate will fall within these limits*, or to quote a maximum value of plant cost which the estimator has 90 per cent (or 95 per cent, or whatever) confidence of not being exceeded.

Simply considering the shape of any probability curve makes it clear that different limits of accuracy are applicable for different confidence levels. And yet investors (clients) do persist in demanding that estimates (and therefore all the work of a study) be carried out to specified levels of accuracy without specified confidence levels.

Except for pre-feasibility work, the engineer is advised to submit his estimate with stated confidence levels corresponding to the stated accuracy limits, based on at least a rudimentary statistical analysis of element cost variation.[4] If project costs eventually fall outside the

[4] There are standard and easily applied software packages available, for example @Risk.

quoted limits, he can at least be justified in saying: 'Oh well, that was the 5 per cent possibility'![5]

9.6 Contingency

Here is another loosely used term. A recommended definition of contingency, which comes from the American Association of Cost Engineers, is:

> A cost element of an estimate to cover a statistical probability of the occurrence of unforeseeable elements of cost within the defined project scope due to a combination of uncertainties, intangibles, and unforeseen/highly unlikely occurrences of future events, based on management decision to assume certain risks (for the occurrence of those events).

This definition is rather a mouthful, and needs to be thought out quite carefully. It may be easier to visualize in terms of the probability curve mentioned above. A contingency is an amount to be added to the estimated cost (assumed here to be the most probable cost, corresponding to the top of the curve, but not necessarily so) to increase the confidence level to an acceptable probability (say 90 per cent) of a cost that will not be exceeded. In this definition, it is implied that a project has a fixed scope, and any elements of approved scope change will be handled as approved variations to the project budget.

This definition is not the universal usage of the word. In some quarters, the 'contingency' is an amount included in the authorized project budget to allow for any variations in scope or any lack of forethought (or whatever) to ensure that the project budget is not exceeded; practically, a margin for error and for future changes – extra money in the bank. This concept has in the past been taken to an extreme in certain plush institutions, where quite large contingencies, even up to 20 per cent, were routinely included in the budget of an authorized project by the executive responsible. This contingency was not passed on to the next level of authority – say the poor old project manager – thus it was possible for the lower level to produce a mediocre performance in terms of budget, while the upper level was a star. This is nice work if you can get it – at the upper level!

---

[5] OK, that is the second thing he will say, after the inevitable condemnation of the project manager!

The extreme example quoted is reflected to some extent in general behaviour. All project managers (and other performers) like to have some extra contingency, often hidden in various innocent-looking expenditure items. Clients and managements are wise to this, and try to eliminate such items from the budget or reduce their value, and as a result the person immediately lower down in the hierarchy finds it harder to demonstrate adequate performance. This is a valid and necessary aspect of management, of promoting efficiency, but the person 'one down in the heap' is well advised to be on the look-out for abuse, the most frequent form of which is a sloppy and potentially one-sided definition of contingency. Most frequently, one-sided practitioners of unreasonable estimate reduction wish to:

- remove from the estimate items which are likely to be needed but are not yet properly defined, and seek to cover these items by minimal allowances labelled as contingency.
- exclude from the contingency any allowance required (by statistical analysis or 'gut feeling'), to change a 50 per cent probability to (say) a 90 per cent probability.[6]

Ultimately, this is a game which is as old as projects, and the project performers have to understand at each level whether they are in a game-playing relationship, and if so how to play it. There are no rules to the game but there is a basic rule for survival of the project performer whose contingency is under attack, which is to avoid sloppy definition of cost components. He should carefully identify all foreseeable cost components and allot appropriate individual allowances, and avoid rolling any such allowances into the contingency. This should be reserved and maintained for unforeseen risks, and for decreasing the probability of overrunning the budget, by increasing confidence levels in a statistically defensible fashion. Such a contingency may be determined by evaluating the possible variation of cost for each component of the estimate, and performing a statistical analysis as outlined in Section 9.5, Accuracy.

As a rule of thumb, to arrive at an estimate which has 90–95 per cent probability of not being exceeded, a contingency of 1.5 times the

[6] Contingency calculation and management are often an exercise of power, rather than mathematics. Some executives use the above techniques to eliminate all real contingency, and then indulge in further power-play by requiring special authorization to access even this bogus contingency.

standard deviation should be added. A standard deviation of about 3 per cent should be achievable for a good estimate, based on:

- well-developed P&I diagrams;
- properly reviewed plant layouts;
- comprehensive equipment data;
- conceptual designs for each discipline;
- individual discipline quantity take-off, including adequate growth allowances; and
- competitive pricing of at least 90 per cent of the cost items by value.

So 4–5 per cent contingency would be required to establish a figure which has a 90–95 per cent confidence level of not being exceeded.

We must emphasize again that this contingency is applicable to a project having exactly defined overall scope. As scope variation is frequently required after project authorization, organizations investing in process plant seldom admit that a plant cost can be evaluated to much better than limits of ±10 per cent until the project is well under way and all designs have been frozen. Put another way, a contingency evaluated as above may be appropriate for a contracting company which is committing to a fixed price for the plant, but the client's project manager probably needs a contingency of another 5 per cent or so to allow for scope variations.

Chapter 10

Plant Layout and Modelling

10.1 Layout

Plant layout is the fundamental conceptual design activity which follows on from the definition of the process and the mechanical equipment. It was noted in the previous chapter, in Section 9.3, Estimating project costs, that a preliminary plant cost could be factorized from mechanical equipment costs with an accuracy expectation of ±25 per cent. Any increased accuracy of costing depends primarily on determination of the plant layout. Of course this may not have to be designed specifically; but the applicability of similar plant layout designs, at least, needs to be decided. The relevance of quoting the order of magnitude of costing confidence is that it gives some idea of the influence of layout design on plant cost.

Following on from the flowsheet, which determines the mechanical equipment requirements, the layout reflects and determines most other plant cost components, and its optimization is obviously critical to the development of a cost-effective plant. It is equally critical to the development of a functional and maintainable plant that is safe and ergonomically acceptable, and can profoundly affect operating costs.

The terminology we are using is that, as was described in the basic sequence planning of engineering work, 'layout' drawings are those which are produced prior to the stage of receiving final equipment details and structural design. The layouts therefore include overall or critical dimensions only, whereas the final 'general arrangements' are those used for controlling and verifying plant construction details. Layouts are intermediate, not construction, drawings. This terminology is not uniformly adopted in the process industry.

Layouts are developed by review and progressive refinement, often considering many alternatives as befits their importance. There is no limit to the number of iterations which may be demanded, so the amount of work to be carried out is often the subject of argument, especially by a demanding client who is paying a lump sum for the work. Layout development belongs squarely in the conceptual phase of the project (or study). If the layouts are not frozen, the project budget and schedule cannot be considered to be final, though obviously there is some scope for debate on what 'frozen' means and what level of detail is included in the freeze.

It is obviously critical that final (frozen) layouts are able to accommodate any variations in equipment size, piping design, etc. that may be required by final detail design. Otherwise there may be at least a major disruption to the design process and schedule while the layout is adjusted, or an unwanted compromise (it may be described more harshly!) in the final plant configuration.

In practice, plant layout development is dependent on experience and design insight, and there is no substitute for having the most experienced practitioners for the job. The following outline is aimed at recognizing the information necessary to advance the layout work, and at reviewing the end result.

Important factors affecting the layout are included within the overall design criteria, including the process design criteria, Appendix 2; the prime importance of these requirements should not be forgotten, even where they are not further elaborated here. The three fundamental considerations of layout are the mechanical equipment, the materials transport systems, and the plant structures, which are mainly designed to suit the first two.

Layout considerations arising from the mechanical equipment, and dependent on the design and operation of the equipment, include the following.

- The method of feeding the equipment and removing product (particularly important for bulk solids processors).
- Equipment access requirements for operation and routine maintenance:
  - starting and stopping
  - observation of operation and of local instruments
  - product inspection and sampling
  - equipment inspection and adjustment
  - lubrication and cleaning
  - catalyst or internals replacement
  - emptying, draining, flushing, venting, making ready for maintenance.

- The means of spillage removal (mainly for plants handling bulk solids).
- Facilities and access for major maintenance, including equipment removal and replacement. These may include permanently installed lifting devices (cranes, hoists, lifting beams, trolleys, etc.) and the suitable configuration of surrounding plant and structures, for example the provision of removable floor and roof sections, and additional flanged connections in surrounding ducts.[1]
- Clearances and special requirements for operational hazards, fire-fighting, and safety requirements.
- Unusual static and dynamic loads, and consequent structural support requirements or isolation of vibrations.
- Noise attenuation.
- Any other requirements or advice from proprietary equipment manufacturers, where applicable. (They should be a prime source of information, comment, and eventual approval, although unfortunately they cannot always be relied on to have good representation.)

It is necessary to check through the above list, at the very least, to ensure that a proposed layout is appropriate for each piece of equipment.

In processes where materials are transported mainly in the fluid phase, such as most oil refinery units, the piping design considerations tend to dictate the layout, whereas for plants processing solids, such as metallurgical reduction plants, the bulk solids handling design is the major consideration.

The engineering of fluid handling systems, including piping and ducting, and their influence on layout, is addressed in Chapter 14. Piping considerations at the layout stage include not only the routing (without creating plant access problems) but also support of the pipes, access to valves, and provision for major expansion loops. Ducting, particularly large diameter duct systems tied into a common stack, can have a crucial influence on layout.

Bulk solids handling is addressed in Chapter 15. Some consideration also needs to be given to the intermediate case of two-phase flow and especially slurry transportation, and this is described in Chapter 16. Bulk solids handling is inherently more complex and less flexible than fluids handling, and has a correspondingly greater impact on layout options. Conceptual engineering of the conveyors and gravity flow

---

[1] It is easy to go too far with this, for example spending money on bolted connections, which will hardly be used in practice (the 'Meccano mentality'), when cutting and welding may be acceptable.

systems (and their elevations) is an inseparable part of layout development, whereas for fluid flow systems there is more scope for employing standardized and modular designs.

The first plant layout operation is to understand and develop the plant arrangements around major pieces of equipment and sub-units of process plant, with particular attention to the shape and size of 'footprint' and the relative positions of main interfaces, especially points at which process materials enter and leave. These sub-units are then combined into an overall block plan development, and when this is acceptable the whole may be further developed into more detailed layouts. Figures 10.1 and 10.2 are examples of finally developed block plan and layout.

During this process, attention is given to the following.

- Reduction of materials transport routes (including those involving major utilities).
- Eliminating or minimizing hazards (for example, avoid proximity of fired heaters to potential sources of LPG leakage, close off areas where rocks may fall, and so on) and allowing adequate fire-fighting access.
- Ergonomics of operation, in particular, adequate operator access.
- The overall site topography with respect to:
  - plant feed source, including major utilities or reagents source where applicable;
  - product destination;
  - location of any intermediate storage;
  - effluent disposal and drainage;
  - existing adjacent plant and facility interfaces;
  - proximity of neighbouring dwellings or places which may be exposed to hazard or suffer nuisance from the plant operation;
  - the prevailing wind direction;
  - utilization of existing site contours to reduce materials handling costs;
  - reduction of excavation and site development costs;
  - overall site access, and construction access and lay-down areas;
  - provision of appropriately positioned space for future plant expansion.
- Plant enclosure standards (roofs, walls, etc.) heating and ventilation, and the impact of environmental conditions.
- Provision and appropriate location of plant service buildings, such as control rooms, operator facilities, laboratories, electrical

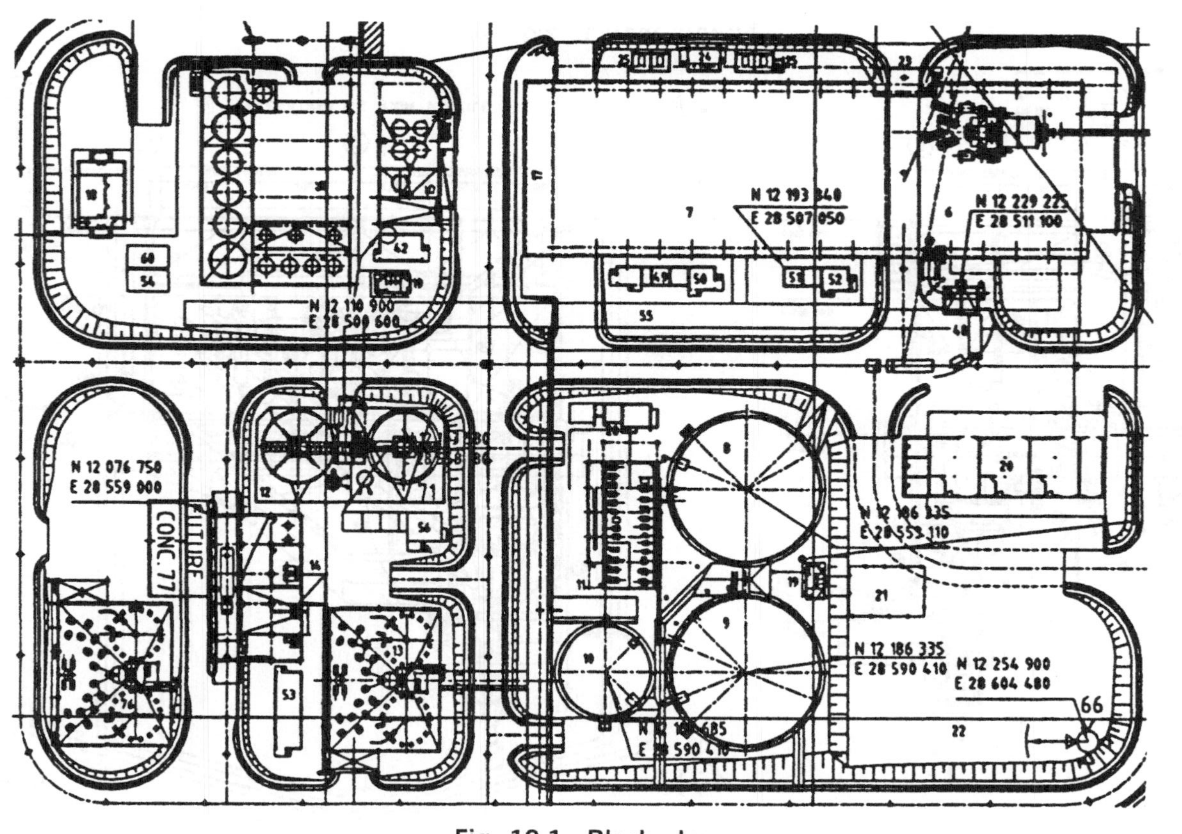

Fig. 10.1 Block plan

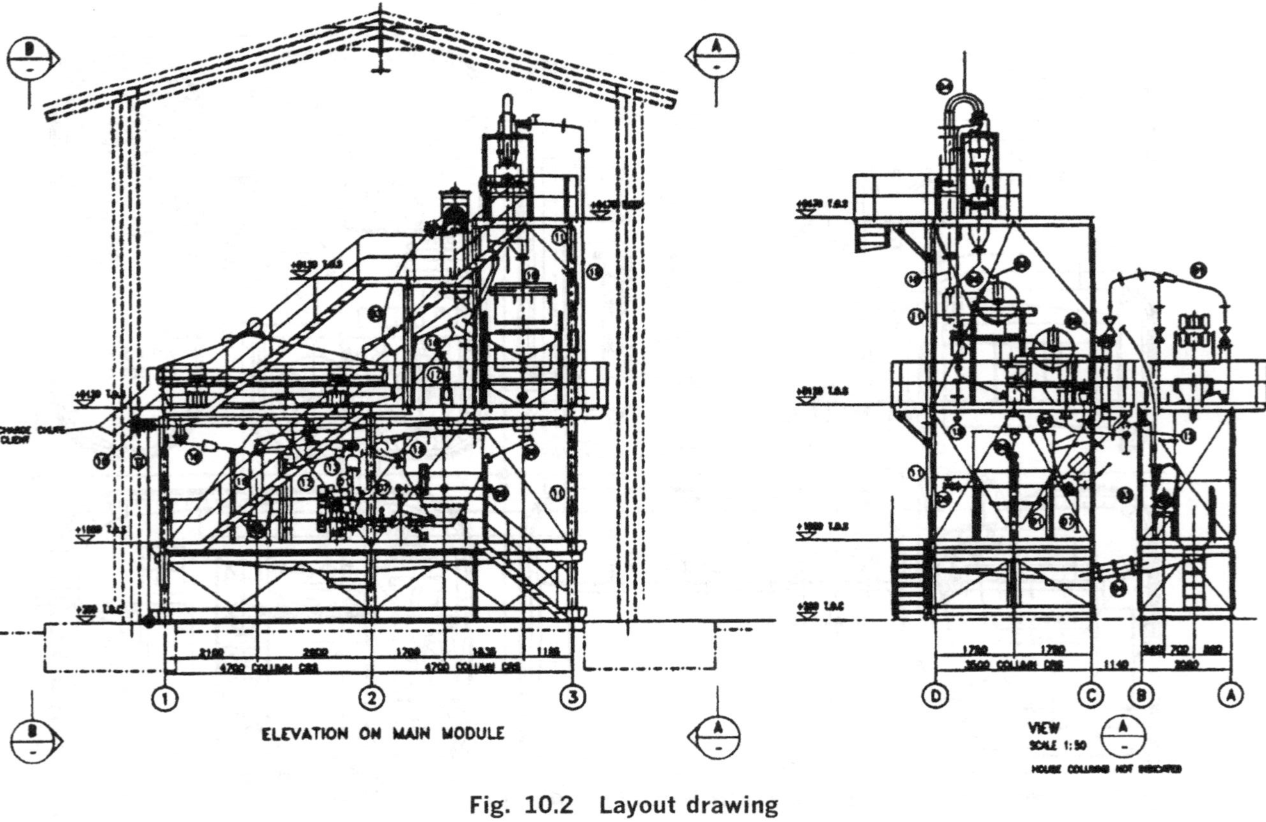

Fig. 10.2 Layout drawing

switchrooms and transformer yards, workshops, and warehouses.
- Internal plant road and access design, and the overall drainage system.
- Routing of piperacks and major ducts, and routing and support of major cableracks.
- Evaluating the possible impact of structural/foundation design criteria and seismic conditions on the proximity of equipment and structures.
- When required, separation of distinct process areas to facilitate shutdown and isolation of each area for maintenance or revamp and, sometimes, to facilitate independent design and construction of those areas.[2]
- Site security.
- Ensuring that adequate space is allowed for design changes that may become necessary as detail design develops.
- Architectural and social requirements relating to plant appearance, and its acceptability to statutory planning authorities.

The plant layout development normally entails an iterative procedure review and redevelopment of the individual plant area designs and footprints until a satisfactorily balanced overall design is achieved. There may be a few major layout options to consider. Selection of the preferred choice includes attention to:

- capital and operational cost;
- overall ergonomics and safety of operation and maintenance;
- constructability and possible modulization to simplify design and construction; and
- space utilization.

It is always of great assistance to visit and review the designs of existing reference plants, both before beginning layout work and when reviewing conceptual layouts. Obviously, such visits are greatly enhanced if operational and maintenance feedback is obtained.

It is vital to check and review layouts rigorously with the use of a formal checklist; the essence of such a list is included above, but this has to be supplemented for the particular processes involved, based on previous experience. Insufficiently comprehensive layout review has resulted in countless cases of inferior or unacceptable plant design.

Once a layout has been approved by all disciplines (including Hazop review where applicable) and frozen, its integrity has to be maintained through the following detail design stages to prevent unauthorized encroachment on discipline space and the compromising of access ways.

---

[2] This subject is discussed further in Chapter 21, The Organization of Work.

This is no easy task, and is exacerbated when design changes – especially process design changes – are made. The use of three-dimensional modelling in such circumstances is strongly recommended.

Apart from individual access points, as the design develops the connectivity of access ways must constantly be reviewed. Ensure that the overall system of access is convenient for plant operation, and that the operators and maintenance personnel are not expected to travel a long way round between adjacent structures.

10.2 Design presentation and modelling

Before going on to the basic materials transport systems design and its layout influence, it is appropriate to enlarge on item 6 (design methodology and standardization) of the project design criteria, presented in Appendix 2, with particular reference to drafting. The objectives of 'drafting' can be met in a number of ways, employing paper drawings, physical models, computer programs, and so on. In order to concentrate on the function rather than the method, we will employ 'drafting' to describe any such process whose objectives are:

- to present a model of the plant and its components in order to facilitate the design and its co-ordination, and to make it possible to judge the acceptability of the design;
- to record the plant design and dimensions; and
- to communicate the design to third parties (that is, outside the design team) for a variety of purposes, such as approval, purchase, construction, or repair.

In this definition, we have separated the drafting function from the design function, which becomes the process of deciding on the plant's configuration, dimensions, manufacture – all its attributes. The draughtsman may be very offended by a supposed implication that all design decisions are made by others (the engineers?) when he knows very well that every time he puts pencil to paper (or clicks the mouse) he makes a decision. He may even feel that his function is to design the plant, while the engineer's function is to take the credit! So let us quickly acknowledge that many, even most, detail design decisions are made by the draughtsman; we are simply splitting the functions of that person into those of design (making design decisions) and drafting (representing those decisions).

The drafting process produces a model (a physical model, a number of drawings, or a computer model) which aids and represents the design, and a communication, which may be the model itself, the model with notes added, drawings, or simply a set of instructions. It should be noted that the communication aspect is an important criterion of acceptability in its own right, and has to take into account the character and needs of the people building the plant.

In Chapter 4, in the discussion on engineering co-ordination, we highlighted the critical importance of the plant model as the instrument by which co-ordination in space is maintained. A model which is clearly and quickly understood by all disciplines (and by other reviewers, such as the client) greatly improves design quality. There is nothing to beat the ease of visualization of a physical model, but unfortunately it is relatively expensive and slows down the design process. Advances made in computer modelling in the last 30 years have taken us to the point where the hardware and software has become almost as economical and effective as the visionaries expected,[3] and most design offices now utilize such systems. If used by adequately skilled staff, they facilitate a process of model creation which matches and becomes the layout, general arrangement, and detail design process (see Fig. 10.3). Mutually consistent drawings are automatically generated. There are also, of course, many ranges of design and analysis software which integrate directly to the modelling software. The trend of integrating associated design and management software has continued to the point where integrated software suites cover almost the entire engineering and management development of a project.[4]

The further development and application of integrated computer-assisted design and drafting is the way ahead, but the choices of methodology to be made for a real, present-day project have to be carefully considered, in particular the need and availability of suitably qualified staff to handle the project's peak load requirement.

Even with the clearest model presentation, there is a need for the ongoing model development to be supervised by dedicated design co-ordinators. The modelling software usually includes clash detection and notification, but it is still necessary to ensure that in conflicts the most important needs prevail and to maintain the integrity of operational and maintenance access and ergonomics. There is also frequently a need for ongoing independent review of the drawings or output communication, from the point of view of their users.

---

[3] Unfortunately, most of the visionary users lost their shirts in the process.

[4] Creating more opportunities for visionary users to lose their shirts in the process! The subject is discussed further in Chapter 28.

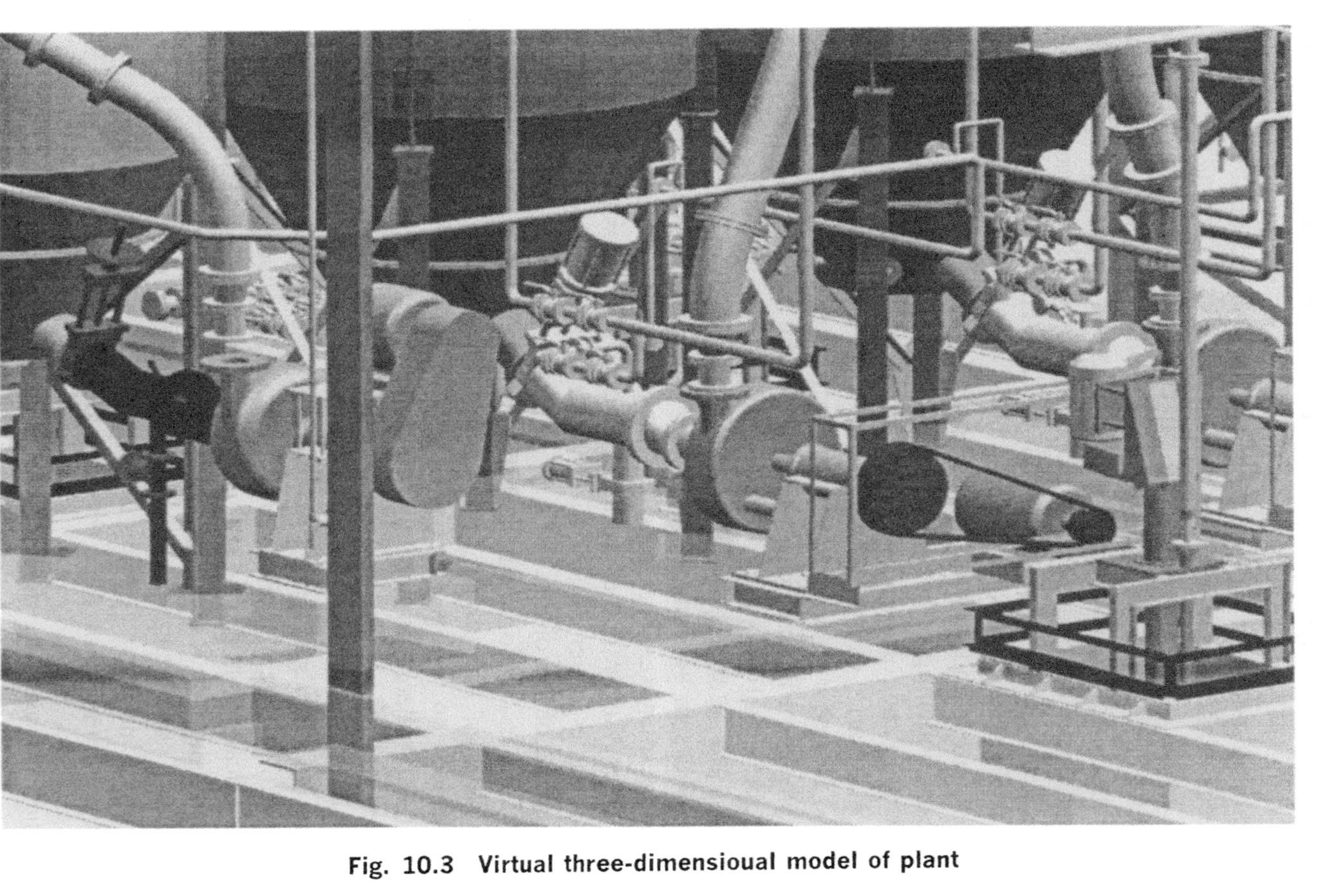

Fig. 10.3 Virtual three-dimensioual model of plant

Fig. 10.4 Autogenous milling plant layout – showing influence of bulk solids handling needs

Chapter 11

Value Engineering and Plant Optimization

Undoubtedly one of the most critical phases of a project or proposed project, the part which can make it be regarded as a star or a mediocrity, a winner or a loser or a non-event, is the creative phase when conceptual designs are developed. There are a few facets to this phase. We have already remarked on the iterative nature of the work done, of successive development, exposition of designs, critical review, and re-work, in search of optimal designs. It is the creative input and innovative thinking that make this process meaningful. In the process plant field, with its heavy dependence on individual plant component reliability, most innovation comes from the use of proven techniques or devices in a different application, rather than the employment of untested new devices. The leading innovators are often people who have a wide knowledge of general industry practice, as well as being innovators by nature. There is still of course room for brainstorming and for the employment of revolutionary thinkers, but since the latter are born rather than made, there is not much to be said about them here.

In the fields of management consultancy and business development, several methods have been proposed to supplement individual experience and brilliance by the use of formalized techniques, whose application is not limited to process plant design. Different systems to promote creative thinking are effective in the cultures of different organizations: each one has to find 'what works for us'. The only general rule is that any design organization that does not promote creative thinking *in the appropriate context*, is obsolescent.

11.1 Value engineering

One system of analysis topromote creative thinking, value engineering, requires special mention. It is often specifically required by clients. We will discuss it briefly, but this should in no way be interpreted as adequate cover of a well-developed subject with ample literature. In its application to the design of process plants, value engineering is essentially the analysis of design by function and utility, to arrive at the most cost-effective design. The analysis requires that a starting-point design exists. The analysis begins with the flowsheet, followed by all facets of the design and its documentation. The questions to ask of each item which incurs a cost or affects costs are:

- What is its function?
- What does it contribute to the performance of the plant and the project as a whole?
- What minimum function and performance are actually necessary? Are there other ways of doing this more cost-effectively?

Evidently, the questions cannot properly be answered without an initial understanding of the plant design criteria in terms of specified overall performance requirements but, aside from these, the content of the plant design criteria should itself be a subject for value engineering to gain the maximum benefits.

One of the prime objectives of value engineering is the elimination of plant features, to arrive at a design where there is 'nothing left to take out'. This obviously has to be tempered against a realistic view of retaining features required for operational contingencies and flexibility of operation, and may make the outcome somewhat subjective and arguable, but there is no substitute for attempting to resolve such issues by quantification. Any facility not strictly required for plant operation should have a quantifiable value, even if the value cannot readily be expressed in cash terms, as may be the case for some considerations of safety (and even this is arguable, as will be discussed in the next chapter).

By the nature of the process, it is apparent that there is no end to the amount of value engineering that can go into a plant design, given that time and resources are available. An engineer who finds himself contracted to 'fully value-engineer' a plant design, within the ambit of a limited budget or schedule, may be in for a sticky time indeed.[1] It is also

[1] If value engineering is required within a fixed-price framework, draw up a procedure of how it will be done and specify what items and documents will be value-engineered, with a specimen output report.

apparent that this process belongs firmly in the conceptual-development phase of a study/project, that is, before the detail design commences.

The following are examples of the application of value engineering and the corresponding benefits.

- Combine separate plant control rooms, and save on both capital and operational costs.
- Eliminate boilers by importing steam from another enterprise that has a surplus.
- Modify process design requirements such that cheaper standard equipment or vessel designs can be utilized.
- Eliminate a storage silo in favour of an open stockpile with tunnel reclamation, the special advantages of silos (protection of material from rain, elimination of dead material, containment of dust, and increased security) being in this case not worth the extra cost.
- Eliminate the agitator from a tank containing slurry; size the tank to be sufficiently agitated by incoming slurry.
- Eliminate pumps; use gravity flow.
- Use a more cost-effective materials transport system; for example, pump solid material as a slurry instead of handling it dry with a succession of belt conveyors when the material is slurried ('re-pulped') anyway at its destination.
- Change a uniquely water-cooled item of equipment to being air-cooled, thereby doing away with the entire plant cooling-water system.
- Eliminate the building enclosing a plant, complete with its overhead cranes, etc.; design the plant for all-weather operation and maintenance by mobile cranes.
- Substitute a control loop for maintaining the level in a fluid circuit by a tank overflow.
- Replace a pressure reduction station (on process fluid or steam) by a turbine which drives an equipment item or generates power, saving more in discounted operating cost than the capital cost.
- Eliminate access platforms and stairs by relocating equipment or a feature that requires access.

Value engineering is more likely to be effective if conducted on an absolutely methodical basis, including at least all of the process flowsheets, plant design criteria, layouts, and cost estimates, obviously skipping over the duplicated items when they come up again. Many of the outputs will seem obvious, once recognized, but may be missed without going through each item on an almost mechanistic basis.

11.2 Plant optimization models

Another important tool for plant optimization is computer modelling of the economic consequences of process configuration and operation alternatives. The systems of analysis which may bear fruit include the following.

- *A basic plant flowsheet-linked economic model*, which simulates the process and calculates both capital and operating cost/revenue implications of different process configurations, capacities, feedstock, product slate, etc. as may be required. Such models are commercially available for plants employing more standardized unit-processes built in high numbers, such as oil refineries, but can of course be built up for any plant.
- *Reliability and maintainability analysis*, based on breakdown, repair, and periodic maintenance statistics obtained for each plant item (or even on sensible guesses from experienced people). In conjunction with economic data, the model is used to predict plant availability and economic performance while taking into account the probability of failure of individual components and the repair time. The model may be used to make optimum choices on the provision of standby equipment, equipment enhancement, operational flexibility enhancement (including optimizing the period between overhauls), maintenance facilities, intermediate tankage or stockpile capacity, and capacity margins for individual equipment items.
- *Economic analysis of plant control options.* This models the dynamic actions of the plant control systems within the overall flowsheet, and evaluates the economic consequences of the design margins employed. Plant design parameters such as pump or conveyor capacities, pressure and temperature ratings, and vessel capacities usually incorporate a design margin to accommodate operational control bands and upset conditions. As a result, the economic optimum design margins can be evaluated. More usually, this analysis is employed to optimize the performance of an existing plant or design, and quite significant capacity increases or overall economic performance enhancement can sometimes be achieved by inexpensive means, for example changing a controller or control setting.

There are a growing number of commercially available software packages for the above, and of course consultants for application and customization. Significant benefits can be achieved but the costs can be high, as can the time required. Many such programs are well-suited for ongoing use as, or within, enterprise management systems and day-to-day plant management tools, and can deliver significant benefits in this capacity. Increasingly, plant owners are realizing that an integrated plan for process and economic modelling, which addresses the whole project and plant lifecycle, should be considered at the stage of project conceptualization and be part of an overall information management policy.

Chapter 12

Hazards, Loss, and Safety

Hazard and risk identification and management have to be addressed throughout the project, but one of the most important times to take stock is before making the decision to go ahead with the project. Social and legal considerations dictate that there are effectively two regimes for addressing risk and hazard: those which should have mainly financial consequences, which we have discussed in Section 9.4, Risk, and those which can result in health impairment, injury, or death, which will be addressed in Section 12.2, Designing for safety. Exposure to commercial risk is regarded as acceptable, and often necessary, as long as the magnitude of risk is regarded as 'reasonable', meaning that the balance of risk and reward is appropriate. Taking chances with safety – including failure to diligently establish whether there are safety hazards – may be regarded as unacceptable and unlawful, but this statement is, as we will see, something of an over-simplification. It begs questions such as 'What is meant by taking a chance?' or 'How safe is "safe"?'

Of course, there is a wide overlap between financial risk management and safety management. Many causes and events impact both, and consequently the systems for problem identification impact both. At a plant operational level, it is convenient to combine the management systems as loss management. With this comes the useful knowledge that an operation which sustains unexpected financial losses is also likely to be unsafe unless corrective action is taken. The incidents that cause financial loss are, at the least, evidence of an uncontrolled environment, and are often mishaps where injury was avoided only by luck. Meticulous investigation of incidents of financial loss often results in corrective action leading to a safer working environment. What follows therefore addresses both financial loss arising out of unwanted incidents and the possibility of injury and damage to health and to the environment.

The project engineer's main responsibilities are the following.

- The identification of hazards, and their possible consequences.
- Designing to eliminate hazards or reduce (mitigate) their consequences to an acceptable level.
- Exercising care in those activities where the engineering team is directly involved in operational activities, which effectively means commissioning.
- Ensuring that the plant operators are warned of the remaining hazards that are inherent in plant operation, and are briefed on how to manage them.
- Addressing hazards caused by engineering project and design activities within operating plants.

Safety during plant construction is an important topic to the project industry, but is outside our present scope.

12.1 Hazard identification

Hazard and operability studies (Hazops) have become the industry norm for process plants of all but relatively simple design, and are generally accepted as being mandatory when there is an inherent process hazard, for instance when the process materials are flammable or toxic. The terminology and methodology were mainly developed within a single industrial organization, ICI. The essence of the technique is that the entire plant design is surveyed systematically, in detail, because it is in the detail that so many hazards lie. Inevitably, this means that the plant design cannot properly be Hazop-ed until the design is practically complete. The reason for discussing the topic at this conceptual stage is to avoid fragmenting it. In fact, it is the perfect example that project engineering has to be learnt and practised in cycles. You cannot properly address the subject of Hazops until you have advanced to a certain degree of definition of plant design, but you constantly need to bear in mind hazard identification and management while the plant is designed. If hazard identification and management is deferred until too late in the design process – a mistake too often made – then the following may arise.

- The impact is likely to be extremely costly as designs and plant already under construction have to be modified, resulting in chaos and delays.

- The Hazop process itself is likely to be compromised. If it is swamped under the volume of problems which surface, the chances of picking up all the problems are significantly decreased. This is analogous to one of the scenarios that a Hazop should bring to light: if a safety device, such as a safety valve, can be used by the plant operators in normal plant operation (rather than taking direct action to prevent an over-pressure), the reliability of the safety system will be unacceptably reduced. In fact, if the Hazop comes up with too many problems, it must be repeated after the modifications have been made.
- The pressure on the project team to gloss over hazards may lead to inadequately safe designs.

So in this chapter, we are dealing with a mindset that must permeate the whole design process – not just a procedure that takes place at the end of the design process.

To carry out a Hazop study, the P&IDs and the plant layout must be available and must have been frozen, that is to say that any future changes must be the subject of a formalized change-control system. Hazop is conducted as a team operation, usually involving at least the process engineer, mechanical engineer, and instrumentation engineer. It is common practice to have Hazop chaired by an expert facilitator from outside the plant design team, and in the case of a project within an existing plant it is essential to include operational people, such as the plant operating supervisor and maintenance engineer. The Hazop is entirely dependent on the knowledge and experience of the participants, and if these are in any way inadequate then suitable consultants must be added to the team.

The essence of the technique is to review every single pipeline, item of plant equipment, and device, and address it with guidewords which usually include:

- None
- More of
- Less of
- Part of
- More than
- Other than
- Reverse.

The possible causes and consequences of the perceived deviations from normal or intended operation are then evaluated and recorded, together with recommended actions to rectify unwanted consequences.

The actions are likely to include both design changes and operational practices, which must be included in plant operation and maintenance manuals, and operator training. A few examples follow.

- 'None' may bring up the possibility that a pump suction may run dry, damaging the pump, and the recommended action may be to install a low-level trip or alarm (or both) on the vessel from which suction is drawn.
- 'More of' relates to quantities or properties of process substances, and leads to consideration of excessive flow, over-pressure, excessive temperature, oversized rocks, etc. with usually fairly clear consequences and preventive action possibilities.
- 'Less of' could lead to recognizing the possibility of inadequate cooling water flowrate to part of the plant when the flow is increased elsewhere, with consequent damage to machinery; the solution could be to install a flow-limitation device on the alternative user, or an alarm.
- 'Part of' is meant to include the possibility that the composition of a material stream may vary, for instance that solids may settle out in a pipeline that has a 'dead leg' under certain operating conditions, with the consequence of blockage and unavailability of that part of the line when it is needed. The recommended actions may include the elimination of the dead leg by employing a recirculation system, redesigning the dead leg to be entirely vertical so that settlement may not occur, or simply prohibiting the operational mode in which the problem can occur.
- 'More than' means more components are present in the system than there should be, for instance the possibility that water may be present in oil introduced into a hot vessel, with the consequence of explosion. The recommended action may be to eliminate the possibility of contamination by water at source, or to detect its presence and initiate a shutdown system, or both.
- 'Reverse' means the opposite of the intended operation, for instance reverse flow.
- 'Other than' looks for any abnormal operating conditions, other than those already prompted.

We said above that it is necessary to look at every single pipeline, item of plant, and device, and herein lies one of the sources of potentially important omission. The definition of 'device' must include anything that can affect the operation of the plant, including power supplies and computers, and evaluating their participation and response in operating deviations.

In conclusion, Hazops take a long time to carry out – at least one hour per mechanical equipment item on the P&I diagram, much more for an inherently hazardous plant – and will result in a number of design modifications. The number of eventual modifications can only be estimated by previous experience, but it should be diminished if the design engineers are experienced, if the plant is similar to a previous construction, and if coarse-scale or checklist-based reviews are made regularly while design progresses. If the project budget and schedule do not include adequate allowance for both the Hazops themselves and the consequent modifications, then the project team is on the first step of a slippery slope which will inevitably result in pressure to skimp on the Hazops and refrain from necessary modifications.

12.2 Designing for safety

Many of the requirements of safety in design should flow naturally from the requirement to design something that is fit for purpose, and checking that the plant will operate as intended under all conditions, as in the Hazop process. There are few, if any, countries that do not have statutory regulations which govern the design, manufacture, and use of some of the common items of plant hazard, for example pressure vessels, structures under load, electrical equipment, and lifting devices, and no attempt will be made to catalogue them here. They are individual to the country of application and, besides, are liable to change. The engineer must simply obtain a copy of the current regulations, and design accordingly.

However, anyone involved with actual design work soon realizes that there are in fact many choices to be made in arriving at a 'safe' design, and that in reality nothing is 100 per cent safe – not even total inactivity, of whose harmful consequences we are all aware! The safety implications of plant design choices are, in principle, quantifiable. The techniques for doing this are usually described as hazard analysis (Hazan).

Hazan seeks to evaluate the probability that an unwanted incident will occur, and the consequences if it does happen. Hazan can only be carried out for identified hazards – it is no substitute for a Hazop, or even for evaluating plant against a checklist of possible problems. Typical failure possibilities that are considered are:

- failure of one or more safety devices;
- operator error in a foreseeable manner, for example failure to respond to an alarm or failure to close a valve;

Fig. 12.1 Operator error in a (?) foreseeable manner

- simultaneous loss of multiple power sources; or
- simultaneous loss of firewater and occurrence of a fire.

History of the operation of certain types of plant, and devices within plants, is clearly the best source of statistical failure information, when available. Sometimes it may be necessary to make an experienced guess or, better, a guess of the upper and lower limits of a figure and consider the consequences of both. For an unwanted incident to take place, usually a number of things have to happen simultaneously.

Firstly, the plant operation has to deviate from the design operational mode, for example the pressure must increase above the intended value, say because of the failure of a pressure controller or because the operator omitted to close a bypass valve. And then, at least one protective device has to fail simultaneously. We can quantify these as follows.

- The demand rate, D, is the frequency at which the hazardous condition (say, the over-pressurization of a pressure vessel) is likely to occur, measured in events/year.
- The fractional dead time, *fdt*, which is the percentage of time when the protective device (for example safety valve) does not operate effectively. This is made up of at least two components:
  - the failure rate, F, of the protective device (the number of times that it becomes ineffective) in events per unit time, say once in a hundred years; and
  - the test or verification interval, T, which could be 2 years for a plant which has a turnaround every 2 years when all safety valves are tested.

If we assume that a failure of the safety valve is equally likely at any time in the 2-year period, then on average the protective device will be dead for 1 year following failure. In general

$$fdt = F \times \frac{T}{2}$$

Given the above, the probability of the unwanted incident occurring will be $D \times fdt$ events per year. The probability can be reduced by installing additional safety devices, say a second safety valve or a pressure switch which deactivates the source of pressure, but each of these will have a fractional dead time, and the incident rate will be reduced to

$$D \times (fdt)_1 \times (fdt)_2 \times \cdots$$

But it will never be zero. And with each extra safety device comes the possibility of spurious action, upsetting plant operation when nothing is wrong.

The second stage of Hazan is to evaluate the probable consequences of the incident. Usually, these are related to the severity of the incident (in the example of the pressure vessel, the maximum pressure experienced in the over-pressure incident). The value chosen may influence both the demand rate and the fractional dead time, so there may be some iteration here. If we continue with the example of the pressure vessel, it is most unlikely that it will explode if subjected to 25 per cent over-pressure – most codes require a hydrostatic test at around 50 per cent over-pressure. There are all sorts of safety margins introduced into the various details of the design but, then again, these are done for reasons which include manufacturing uncertainty and service uncertainty (say, the inclusion of a corrosion allowance). It becomes clear that the pressure at which a vessel will fail is itself a matter of statistical probability. In practice, it is frequently the case that an over-pressurized vessel or pipe fails at a flanged joint, and the consequences (to people) are not so severe unless the contents are flammable or toxic. For the given example, let us suppose that we can expect the vessel to fail at a pressure of 170 per cent of the rated pressure, that we have used this pressure in evaluating the demand rate, and that a similar failure in the past has resulted in two fatalities.

Having done our best to quantify that there is, say, a 0.1 per cent chance that a failure will occur in 100 years, and as a result an average of two fatalities may arise in the ensuing explosion and fire, we come to the final stage of evaluation, which is to decide whether that is acceptable. This is commonly done by comparison with statistics that relate to normal hazardous activities that most people are prepared to engage in, like road or air travel. The designer may be able to demonstrate by this means that the various features of hazard around a plant create an environment where persons may be exposed to (say) less than a tenth of the probability of injury than when travelling by road. It may be argued by such means that the standard of safety is adequate (or, of course, inadequate if the results come out worse than road travel).

It has to be stressed that, in the event of an accident, there is no guarantee that such quantified evaluation of hazard has a certainty of being accepted as proof of reasonable behaviour by the plant designer. In practice, it is nearly always possible to make design decisions and choices by adopting a suitable design code that obviate the need for hazard analysis unless that is specifically required by regulators (invariably, in addition to the adoption of proven design codes of practice).

However, the practice of performing hazard analysis is most enlightening for the plant designer, if only to enhance the understanding of practical reality, which does not easily flow from blind obedience to codes.

On the general subject of conducting Hazops and identifying and quantifying hazards, the reader is referred to the text *Hazop and Hazan – Identifying and Assessing Process Industry Hazards,* by Trevor Kletz, published by the Institution of Chemical Engineers. This is a very readable further introduction to the subject. More comprehensively, the website of the Institution of Chemical Engineers, www.icheme.org, lists at least 50 (mainly specialist) publications on the subject.

In conclusion, to design a safe plant the following are required.

- Ascertain and incorporate the relevant requirements and regulations of the Health and Safety Executive and other statutory bodies within the country of the plant site.
- Check for regulations or permitting requirements concerning emissions and waste disposal. An environmental impact study/plan is normally required, and is likely to include some recommendations which will affect the design. Do not omit to contact and ascertain the requirements of the local fire office. Check also whether there are, or are likely to be, any design requirements from the owner's fire and other hazard insurers, in case the owner has omitted to mention them.
- Adopt industry codes of practice for the plant as a whole, where these are available. The 'process design package' mentioned in Chapter 2 ought to have a comprehensive reference to applicable codes of practice and design requirements that are appropriate to the process, and, if not, these should be demanded. The subjects addressed should include:

- the sizing and design of pressure-relief systems;
- requirements for redundancy of safety devices and for fail-safe systems;
- requirements for on- and off-line testing of safety devices;
- requirements for and frequency of plant equipment inspection;
- design for safe entry into enclosed spaces;
- the classification of hazardous areas;
- fire prevention and protection;
- plant layout for safety, including minimum clearances between adjacent units, and proximity to places of public access and dwellings.

Further information on appropriate codes of practice (and good practice in general) can be obtained from the Institution of Chemical Engineers website mentioned above, including the comprehensive *Guidelines for Engineering Design for Process Safety.* For hydrocarbons work, visit the website of the American Petroleum Institute or, more easily, download their publications catalogues.

- The project design criteria (see Appendix 2) should include at least the design codes to be employed for (where applicable):
  - vessels under pressure (including plastic vessels, if utilized);
  - pressure pipework;
  - fired heaters and boilers;
  - non-pressure vessels which contain hazardous substances;
  - safety of mechanical handling devices and guarding of machinery (see also the final section of Chapter 15, Bulk Solids Transport);
  - electrical safety, including identification of hazardous areas;
  - lifting devices (including elevators).
- Hazops are required, as outlined above.
- The plant layout must be comprehensively reviewed – this can be incorporated in the Hazop – to ensure that the design is ergonomically friendly and that adequate escape routes are provided.
- All hazards must be described in plant operation and maintenance manuals, together with the correct operational practices to overcome them and keep the plant safe.

12.3 Commissioning

Refer to the notes included at the end of Chapter 24, Commissioning.

12.4 Plant modifications

Making modifications or additions to existing plant requires special safety considerations. One of the most infamous process industry accidents of the twentieth century, the Flixborough disaster at a plant in England, was caused by an insufficiently analysed modification to an existing plant. Any engineer who works in an oil refinery or petrochemical plant is very quickly made aware that any plant modification whatsoever is regarded with great suspicion. In most such environments, a regime is maintained under which any change whatever – even, say, a different type of gasket, because the usual gasket is not available – requires special authorization. A bureaucracy is set up with the intention of forcing people to think critically, and ensuring that changes are reviewed from all operational and engineering viewpoints.

Here are some points to bear in mind, for the uninitiated engineer working in such plants.

- Hazop is required for all but the most minor plant alterations and additions – certainly any work that affects the connectivity of process flows ('jumpovers') or affects the operation of a safety device.
- Permits-to-work must be obtained from the operation supervisor for any work in areas within, connected to, or adjacent to existing plant (including decommissioned plant). The permit must specify at least:
  - what work may be done (generally divided into hot or cold work, depending on whether it is permissible to introduce sources of ignition, such as welding);
  - when and where the work may be done;
  - what precautions must be taken for (where applicable) the detection and elimination of harmful or flammable substances, making safe the operational plant, isolation of pipelines and power supplies, access to enclosed spaces, fire-fighting, and gas inhalation;
  - any other requirements for safety, including personal protective equipment; (normally, these will be covered by mandatory plant safety regulations);
  - what operational supervision is required;
  - the duration and renewal requirements for the permit.

  When work has been completed, there needs to be a formal check and acceptance of the work by the plant operators, and the permits must be cancelled.
- Underground services, and in particular buried electrical cables, must be identified before any excavation is made.

Fourth Cycle
Engineering Development
and Detail

Chapter 13

Specification, Selection, and Purchase

13.1 Procurement

As we have seen from the sequential planning of engineering work, not only are the project designs and their practical implementation connected through the process of procurement, but also the flow of information (both technical and commercial) back from vendors is essential to keep the design process on track. There can be no possibility of a successful project engineering effort without an equally successful procurement effort; the two functions must be integrated from the basic planning stage through to project completion. In Chapter 6 we discussed some of the broad issues of procurement, in the following the subject will be addressed at an operational level.

The engineering task that initiates the procurement process is the defining of what has to be procured, by means of specifications, data sheets, drawings, and work-execution plans, and this definition has to be geared towards the capabilities of the suppliers, based on a considered strategy and a knowledge of the marketplace. There remains the work of:

- drawing up conditions of purchase and contract;
- packaging the commercial and technical content, and ensuring that all aspects of the proposed agreements are covered;
- soliciting bids;
- receiving the bids systematically, and assessing them technically and commercially;
- negotiating and finalizing agreements with selected bidders; and
- following up to ensure that the correct goods and services are provided according to schedule, and that payment is made as agreed.

Invariably in large projects and organizations, the part of the work listed above (except for the technical content) is performed by commercially trained staff under separate management, thus providing appropriate skills and focus to the tasks. The separation of functions creates an important interface, and the following discussion relates mainly to its management.

For successful teamwork, it needs to be understood that the separation of roles is artificial: every important procurement document, decision, and priority is shared between the engineering and commercial functions. If the two functions are too far separated, topographically, by over-formal communication or by over-rigid interface procedures, there will be a price to pay in terms of efficiency and speed. However, interfacing is improved by using well-thought-out procedures, as long as it is understood that these are a tool to promote teamwork and not a substitute for it.

The first mutual need is to plan the work (and the procedures to do the work) in a way which provides the best balance between technical and commercial issues, as discussed in Chapter 6. The following are some of the points to be resolved:

1. Identification of the orders and contracts to be placed. The lists of equipment and bulk materials to be purchased and erected on site are obviously the basic starting point, but it has to be decided how these will be 'packaged' into orders and contracts. The following objectives need to be considered in reaching the best compromise.
   - Best commercial policy. Usually, the combination of items within a package which is judged to be most appealing to the marketplace: large enough to be attractive but not too large for the capacity of target suppliers; not so small that procurement management costs become excessive; structured to include mainly goods and services within the product range of individual suppliers.
   - Technical standardization. All like items of proprietary design should come from a single supplier if possible (and if a good commercial deal can be struck), both for ease of project execution and to facilitate plant maintenance.
   - Schedule needs. Three aspects need to be considered for each item in a package: when is information available to purchase it, when is the item required on site, and, in the case of items of proprietary design, when are the vendor drawings and other interface information needed?

- In the case of site services, such as site erection: the overall project construction strategy, including manageability of contractors, access, desirability of maintaining competition for extra work, and infrastructural limitations.

2. The format of commercial documents, such as enquiries, orders, and contracts; and, arising from that, how the technical documents, such as specifications and schedules, will dovetail into the commercial documentation, such as conditions of purchase and forms of tender, without any conflict and without the need for revision. Subjects which need to be interfaced include:
   - vendor document requirements,[1] both with bids and during the execution of orders, including their quantity, format, quality, timing, and the structuring of sanctions in the case of late or inadequate information;
   - inspection procedures;
   - concession procedures;[2]
   - packaging, marking, and forwarding requirements, including definition of point of delivery;
   - commercial conditions corresponding to performance testing and inadequate performance;
   - requirements (operational and commercial) regarding site attendance by equipment vendors;
   - spare parts ordering system, including commissioning spares; and
   - procedure for project communication with vendor (two-way: the main need is to promote quick technical communication, while creating adequate records and maintaining the commercial controls).

3. Identification of which items will be handled as purchase orders, and which as contracts. This chapter is aimed more at purchase orders than at contracts, for which some additional considerations apply; these are addressed in Chapter 23.

---

[1] Checklist: drawings (preliminary, for comment and approval, certified for construction); technical data, including all interfacing data, such as structural loading, electric motor details, and instrumentation signals and connections; manufacturing schedule; inspection and test certificates; installation, operating, and maintenance manuals; lubrication schedules; spare parts guides and price lists; and packing lists.

[2] That is to say, requests for a relaxation from specification or approved vendor drawing/data detail, arising because of a manufacturing or sub-supplier problem.

4. The time schedule for interfacing of activities.

Organizations which have a responsibility split between engineering and procurement invariably need to use a formal interface document, the requisition, that is effectively an instruction to purchase which provides the relevant technical information. Like all forms, the requisition form is essentially a checklist, and its format conforms to the interface agreement reached in accordance with the above.

13.2 Specification

The word 'specification' has different meanings in different contexts, especially in legal and patent usage. In engineering projects it means a detailed description of the design and/or construction and/or performance of an item. The item can be an entire process plant, or one or a group of pieces of equipment, activities, designs, or bulk commodities. The usage of specifications is as old as the performance of projects, and is the epitome of the statement made in the preceding review on project management, Chapter 3, 'Plan the work then work the plan'. First, you decide exactly what you want (the specification), then you do it or get it.

The project design criteria document is thus the most basic of the project specifications and, below that, any individual discipline detailed design criteria and working practices, developed to guide the work of the project in a manner acceptable to the overall objectives and/or the client. There is some overlap in the meanings of 'specification' and 'procedure'. Generally, 'specification' defines the product of an activity, whereas 'procedure' defines how to get there. A procedure may be part of a specification, and vice versa.

Here we are discussing procurement, and the specifications therefore address the interface between the project organization and the supplier/sub-contractor, in order to describe exactly what is required. In the course of procurement, the specification is used in two contexts: firstly, as the document which makes a competitive bidding process possible and, secondly, as a reference for the work performance, describing the products or services that are to be provided, and the standards of acceptability.

13.2.1 Specification as a document for bidding

Looking at the first context, in a competitive procurement environment our objective is normally to procure the most cost-effective product, that is, the product that fulfils the required functions, with the required reliability and durability, and at the minimum cost. The latter may be adjusted to reflect the evaluated worth of performance-related features, such as power consumption or maintenance costs, as discussed in Chapter 8.

In order to obtain the best value, it seems sensible to produce a specification which reflects the minimum requirements that are acceptable, thus enabling maximum competition and possibly innovative solutions. In effect, this is a value-engineering approach. The specification is aimed at identification of the essential requirements for performance of the item, 'performance' meaning the definition of function and how well, how economically, and how reliably the function must be performed. Any non-essential requirements are identified and deleted.

This is one of the basic tenets of writing specifications, but limitations to the practice include the following.

- If basic design choices are not made when detailed engineering is in progress, it becomes impossible to complete the work. For instance, with regard to equipment selection, once the layout has been frozen (based on certain equipment types) it usually causes a lot of disruption to change to other equipment types. The effect on the layout, the occupation of space, the supporting steelwork design, and on other disciplines just becomes too great and the potential benefits are far outweighed by the cost of change.
- As regards bulk items (such as pipe fittings), standardization and all the associated benefits will be lost if only functional requirements are considered each time an order is made.
- Referring back to the discussion in Section 8.2 on lifecycle considerations, we remarked on the difficulty of confirming maintenance cost and reliability claims made by equipment suppliers. We concluded that it was often preferable to specify features which were known to offer enhanced performance in these respects rather than rely on competitive bidding to produce an optimal design.

The preparation of specifications is therefore a compromise between getting the best value by restricting the requirements to the essential performance of the item, and the sometimes conflicting needs of design convergence, standardization, and inclusion of proven essential features. All of these factors are important.

To some degree the compromise can be reached by using two stages of specification. In the initial or study stage the specifications may be almost exclusively performance-based, and serve to establish the basic equipment choices, standards, and essential features. In the second (or project) stage, the permissible variations may be limited in accordance with the choices made in the first stage. In practice, the stages are seldom completely separated, and the engineer is left with some continuing need to compromise; however, the performance of the work is enhanced if the two-stage separation, the process of design convergence, is seen as an objective.

The commercial goals of getting the best value, and of encouraging competition, are enhanced by specifying items available from existing standard products rather than special-purpose items. The definition of performance and any other features should as far as possible be referenced to national or international standards that are familiar in the suppliers' market. Similarly, and especially for lower-value items, the specification should be as simple to read and interpret as possible. Length, complexity, and non-standard requirements are all features which can be expected to cause less competition and higher prices.

Standardization within the project, ease of reference to the user, and economy of engineering time are all assisted by making up specification appendices for standard project information rather than repeating these in the body of each specification. There may be a number of appendices, for example:

- site conditions (location, weather, altitude, atmospheric corrosion, seismic design), utilities availability, and the applicable utility design parameters (such as cooling water temperature and pressure range, instrumentation and control interfaces, power supply, and compressed air details);
- basic requirements for proprietary equipment – engineering standards such as system of dimensions, design life of machines and components, language to be used in nameplates and manuals, equipment tagging instructions, and site lubricant standards;
- for each discipline: component standardization such as types of nuts and bolts, seal, grease-nipple, and switchgear;
- vendor documentation requirements;
- shop inspection, packaging, and despatch;
- painting and protective coatings.

The body of the specification is then confined to information which is peculiar to the item in question. The use of such appendices has to be limited to procured items of appropriate size, otherwise it is found that relatively minor items may end up with several appendices which are for the most part irrelevant, thus defeating the objectives of the previous paragraph.

13.2.2 Specification as a reference for work performance

The second stage of the utilization of specifications is as a reference for work performance and acceptability. It is usually the case that, either in his bid or in subsequent negotiations, the supplier offers various desirable features which were not specified, or features which differ from the specification but are considered acceptable. For the second use of the specification, as a reference document for the acceptability of the item and its manufacture, it is necessary to update the specification accordingly 'for purchase'. It should be noted that at this stage there is no longer any value-engineering purpose served by seeking to focus on functional needs – it is a matter of ensuring that what is delivered is what has been agreed on.

When updating the specification, it is recommended that the engineer use wording which reflects responsibility for the features proposed by the supplier, the point being that if the supplier in his offer warrants the acceptability of these features, he should retain responsibility for his claims. In the event that his claims prove to be false, he should be held accountable. This is easily achieved by modifying the specification for purchase, where appropriate, by using wording such as 'features offered with supplier' or, in the case of a performance data sheet, by clearly noting on the form which entries are the supplier's. What should be avoided at all costs is to substitute the performance-based specification with details of the supplier's offer, unless the engineer is consciously taking full responsibility for ensuring that what is offered will perform acceptably.

13.3 The selection of equipment

'Equipment' is almost invariably purchased from suppliers who provide their own designs of items to fulfil a specified duty. Thus the project engineer's role is to specify and select, rather than to design, as is usually the case for structural steelwork, civils, and piping. Consequently there are some special considerations to be discussed.

In general, equipment is developed, rather than designed, for an application. That is to say, designs for a new device 'off the drawing board' are seldom suitable for incorporation within a plant without field-testing, during the course of which unexpected problems arise and modifications are found to be necessary; or, of course, the design may be found to be unsuitable. The duration of 'field-testing' must ultimately be compatible with the length of service and reliability expected of the project application, usually a matter of years.

It logically follows that, except for rather simple applications, no engineer can be expected to pronounce an equipment design as being fit for purpose simply by studying drawings and data, even a non-working model. The reviewing engineer is in no better position than (and usually at a considerable disadvantage to) the original designer producing an untested design. So, even with the most thorough design review by the most experienced engineers, it is usually unacceptable to select equipment merely on the basis of its design; the track record for the application also has to be validated.

Engineers tend to be constructively minded people who want to build something new, and want to do it themselves. The optimal behaviour in the task of equipment procurement therefore goes against the grain for many engineers; they want to take over some of the functions and responsibilities of the equipment vendor, and this is especially the case for younger engineers. Here are some suggested guidelines for novices approaching this task; they may be 'obvious' to more experienced engineers, but are too important to risk omission.

- Both commercial and technical needs demand a clear distinction of technical responsibility between the purchaser and the vendor. Commercially, the precise definition of the vendor's responsibility assists the comparison of offers and the manageability of contracts, and is appreciated by vendors who wish to market quality. Definition of responsibility is always fundamental to quality management.
- Unless there are overriding factors of specific technical experience on the part of the purchaser, the best split of responsibilities (which should be regarded as the 'default' case) is that the purchaser is responsible for:
  - describing the application (including the process and local environment);
  - defining the performance; and
  - stating minimum[3] constructional requirements (over which the

[3] *Minimum* implies that the vendor is still responsible for the *actual* features employed, provided that they are equal to or better than the minimum.

performance requirements take precedence in the event of conflict).

The responsibility of the vendor is to design and manufacture equipment which complies with the specification and is fit for the purpose and environment, as described, and as known by the vendor in terms of his previous experience.

- When dealing with the 'approval' of the designs and data submitted by vendors, it is necessary to maintain the responsibilities of each party. The prime responsibility of the purchasing engineer is to satisfy himself as well as he can that the equipment offered complies with the specification. This is best handled by asking questions, if necessary *ad nauseam*, until satisfactory responses are obtained; or, if satisfactory responses cannot be obtained, by rejecting the equipment. Any design modification or detail proposed by the purchasing engineer is a suggestion, not a request or a requirement.

Most standard commercial conditions are based on similar precepts to the above. Some practitioners take the role definition a step further by avoiding use of the word 'approval' altogether, and using 'review', so as to emphasize that there is no responsibility shift. Others argue that this is going too far – that the implied responsibilities can be made quite clear by the conditions of purchase, and that the right of approval is essential.

13.4 Technical appraisal of equipment bids

Following the issue of specifications, the second major responsibility of the project engineer is the technical appraisal of bids for the equipment item. The objectives of this process are the following.

- To check for compliance with the specification.
- To evaluate the relative worth of features which will impact on plant operational costs, specifically:
  - energy and utilities consumption cost;
  - operator attendance cost;
  - maintenance costs (over the life of the plant, including eventual replacement costs if required); and
  - costs arising out of scheduled and unscheduled item unavailability during repair and maintenance; this includes the consequential cost of equipment failure.

Needless to say, the value of these features has to be discounted, as outlined in Chapter 8; and as mentioned in that chapter, concerns raised about reliability are liable to result in rejection rather than evaluation of associated costs.

- To evaluate the relative worth of any capital cost impacts which do not show up in the item prices. These may for instance include costs of transport, erection and commissioning, associated foundations, electrical reticulation, instrumentation and control gear, and pipework. Normally the parallel commercial appraisal, excluded from this scope, will include such factors as impact of terms of payment, duties and taxes, foreign exchange, import/export permit availability, and terms of contract. The interface between the engineering and commercial appraisals needs to be co-ordinated.
- To evaluate the worth of any other features offered by the vendor, over and above those specified, such as ability to exceed the specified performance.

The appraisal is customarily presented in columns of comparison of corresponding features. The work of preparing this presentation is very significantly simplified if the tabulation format is thought out in advance and presented with the enquiry as a requirement for bidders to complete.

Some bidders may be found to submit only partial information by oversight or confusion, or due to genuine lack of time, and there is usually a commercial process to permit the provision of additional technical information after bid submission. Other bidders may however submit inadequate technical information as a matter of policy, their aim being to promote their ability to negotiate and to vary technical details (and the associated costs) according to their subsequent assessment of their negotiating position. As a general rule, project engineers are advised simply to reject bids with inadequate or contradictory information, provided of course that they are in a position to do so in terms of the adequacy of competing offers. Failure to do so often leads to a drawn-out bid appraisal process which is highly damaging to the project schedule, besides being unfair to the other bidders, who may be expected to respond negatively in the future.

The first appraisal criterion listed above (compliance with specification) should be self-explanatory. The second group (operational costs) was discussed in Chapter 8; the most difficult aspect to assess is usually reliability and its impact, followed by maintenance costs. Sometimes, the only way to approach these subjects is to take the view that the vendor is offering a pile of junk, unless he can prove otherwise by evidence of direct and comparable experience and reputation. The experienced appraising engineer is usually able to differentiate features associated with 'ruggedness' – an ability to withstand a certain amount of abuse – which is essential in many applications.

As regards the third group (impact on other capital cost components), there are a couple of aspects which are worthy of elaboration. The items appear on this list because they represent activities and costs for the purchaser which are affected by design decisions and information provided by the vendor. Some of the items listed – foundations, erection, commissioning procedures – may be made to be unnecessarily expensive by vendor design decisions, for example by the imposition of relatively demanding tolerances on foundation interfaces. It may be that certain vendors require unachievable tolerances in erection, and unreasonable maintenance practices, as a means of pre-empting the enforcement of guarantee obligations, or even as a means of gaining extra income from the prolonged site presence of very expensive vendor technicians. User-friendly features, such as adjustment facilities within intermediate soleplates, or shaft-coupling arrangements which facilitate alignment, may be of significant value when constructing (and maintaining) machinery.

13.5 Inspection and quality control

'Inspection' and 'quality control' are two overlapping activities. In a general engineering context, 'inspection' includes both quality and quantity examination and verification, and indeed the exercise of due diligence in examining anything of importance, which in the case of a project may include the site conditions or the wording of a contract. 'Quality control' includes observation and record of product quality, analysis and observation of the production process (including all aspects of engineering and management), surveillance of the actions taken to maintain quality within acceptable limits, and rejection of non-conforming work.

The stages of quality control prior to commissioning can be divided into three: engineering, procured items, and fabrication and construction work. Engineering quality control has been addressed in previous chapters. In the following, our focus is on procured items and shop fabrication, and we will in general use the term 'inspection' loosely to embrace both activities of inspection and quality control, i.e. the observation and the feedback loop.

In the pursuit of providing a quality product (that is, a product having *any* specified quality level), inspection is arguably just as important as design and production. There is no gain in designing features which do not materialize in practice. Product inspection, in all its forms of quality, quantity, and general conformance checking, has long been recognized as an integral part of any productive activity or commercial transaction. One of the fundamental laws of purchase of goods that already exist, *caveat emptor*, 'let the buyer beware', stresses the presumed negligence and lack of recourse of the buyer who fails adequately to inspect.[4] The tradition in engineering is generally to assume that errors and unacceptable work will inevitably follow poor quality control, both in design and construction. All important work must be checked, and corrected if necessary.

In the present limited context, our aim is to establish what inspection, and how much inspection, is required when procuring items for a process plant. Evidently this is very similar to the question of how much engineering work to do; too little inspection is likely to result in unacceptable quality, while increasing the inspection expenditure will ultimately become uneconomic as the costs exceed the benefits. Theoretically, the optimum is reached when the marginal increase of inspection costs equals the marginal decrease in direct field costs and other savings arising out of the additional inspection. 'Other savings' may flow from reduced site erection problems, reduced commissioning time, increased plant availability, reduced maintenance costs, and so on. Clearly, the optimum amount of inspection for an item is influenced by the quality of the system by which it is produced, and by the consequences of a flawed product being accepted.

Some possible consequences of poor product quality are unacceptable, and the corresponding items should be identified by review of criticality if they have not already been identified by the process technology package or by statutory regulations. Clearly these items *must* be inspected to certain standards, and the engineers should ensure that the items are known and the standards are defined. Pressure vessels are an example.

A process plant is finally inspected when it is handed over as a complete entity, usually prior to commissioning. However, this is not a practical nor an economic time to concentrate inspection activities;

---

[4] This is not universally applicable; not for instance when there are other overriding laws, or when the purchaser has 'qualified-out' his obligation to inspect, or where there is a warranty from the seller.

impractical because most of the critical features become inaccessible during construction, and uneconomic because of the cost of delays and of dismantling to carry out rectification. Activities and components have to be inspected before their incorporation into the plant, so that ideally the final inspection should be a verification that all of the planned, preceding inspections have been completed and deficiencies remedied.

Inspection can be classified according to who is responsible for the activity; there are usually many levels to this chain. Typically, there may be the inspectors of the project management and engineering team, the quality department of the procured item suppliers and contractors (and, separately, their factory production workers, using inspection as a production control tool), and a chain of sub-component and raw material suppliers and their quality departments.

There may also be superimposed inspection from the plant owner or third-party bodies, sometimes acting within statutory or licensing agency authority. We will disregard the latter as there is usually little choice to be exercised in how their input is applied, and because they relieve the project team of no responsibility. There is however an additional burden in managing the interface with such agencies and of ensuring that documentation and work practices follow their requirements, which should not be omitted when drawing up inspection budgets.

For most project inspection activities, the interests of economy and efficiency are best served by utilizing the internal quality control systems of the suppliers/contractors, and monitoring the control system rather than the end product. Thus standard conditions of purchase should specify the quality management systems required of suppliers/contractors, and the facilities for surveillance of the systems. End-product inspection, if carried out by the project team inspectors, should essentially be regarded as a tool for monitoring the effectiveness of the quality control system, and if failures are detected both the system and the end product have to be upgraded. For instance, it is common practice, when specifying sample radiometric inspection of welding work, to require that if detected flaws exceed a certain percentage of samples inspected, the percentage of welds to be inspected, both retroactively and in the future, increases to perhaps 100 per cent. This may continue until such time that the pass rate improves sufficiently, and is all at the contractor's expense.

The practice of purchasing according to specification, which the supplier is responsible for meeting, allows the buyer to escape the responsibilities of *caveat emptor.*[5] The buyer's inspector must be careful to assume no responsibility by the act of inspection, exactly according to the reservations of the buyer's engineer in the case of design approval.

There is a simple saying, 'You cannot inspect quality into a product', which is in fact a half-truth that causes some misconceptions. The intent of the saying is that the quality of a product is primarily a result of the production process. There are several ways that inspection augments quality or, conversely, that inadequate inspection leads to quality deterioration.

- At the most fundamental level, *any form of quality control is impossible without inspection.* Control of any form depends on a closed loop of inspection and correction.
- As in quantum physics, *observing a process changes the process.* The performer who is watched jacks up his act, and if he gets into a situation where he cannot cope, any part of his workload which is not subject to inspection tends to suffer first.
- In the case of 100 per cent-inspected product, *once all the rejects have been separated the increase in quality rating of the remainder is obvious.* Even the rejects can be increased in quality by assigning a new or restricted duty to their usage, and for this revised application the quality may be regarded as adequate.
- In conjunction with adequate record-keeping, inspection facilitates the identification and rectification of problems which may occur over the lifetime service of a component or group of components. This is especially the case where such problems may arise out of design or specification inadequacy – the existence of the inspection and test documentation has ongoing potential value, rated as essential where the consequences of failure are severe.

13.6 Planning inspection work

Enough of generalities; how do we plan our inspection work on a real project? Inspection activities should be classified according to the aims

[5] We have not discussed the use of second-hand equipment, which is sometimes the key to project viability. One of the main consequences of buying used equipment is that the vendor cannot easily give a serious guarantee, except by completely stripping and refurbishing the equipment at substantial cost (which is not known in advance). Thus the onus of establishing fitness for purpose falls partially or wholly on the buyer, who must tread warily. The situation is better if the used equipment belongs to the plant owner, who is able to confirm its operational and maintenance history.

and needs relating to groups of items. There are many ways in which this is practised; here is one example.

Group 1. These are critical items where the consequences of failure are severe. An exercise of due diligence is necessary; that is, the intensity of inspection should at least equal customary practice for comparable work. Typically, all work should be carried out in accordance with an appropriate code or standard, with an independent inspector to certify compliance in accordance with an individual quality plan per item.

Group 2. These are items where compliance is required by statute, for example pressure vessels. The applicable regulations dictate the minimum requirements. If these items are also group 1 items, above-statutory-minimum requirements may be needed.

Group 3. These are items where conditions of contract or commercial policy dictate the levels of inspection, for example, in case the inspection is linked to a final payment. There are no short-cuts here.

Group 4. These are the remaining items, where inspection is carried out on a basis of economic evaluation,[6] considering the selected supplier's reputation and facilities, the availability and credibility of conformance certificates, the consequences of accepting non-compliant items, and the cost of inspection.

For management purposes, the decisions on inspection intensity may be classified into a few standard 'levels', for example no shop inspection, random inspection of end product, and individual item quality plan.

Notwithstanding the above, most project inspection budgets are under severe pressure, as rather a cynical view is taken on the value of inspection, generally the result of unthinking criticism when defective products pass inspection, without any understanding of the economic impossibility of 100 per cent inspection. Possibly the most effective countermeasure to the tendency to slash inspection budgets is to work into conditions of purchase the requirement that suppliers and contractors are obliged to pay for appropriate levels of additional inspection (by third parties) in the case of failure of sample inspections. This may also be the fairest and most effective option. Too often, in modern practice, the quality supplier loses on price to those whose quality control is purely cosmetic, and who ultimately (when quality failures are noticed, too late) inflate the purchaser's inspection costs while destroying his inspector's reputation.

[6] In other words, the optimum input should be determined on the basis of diminishing-return theory, as in Chapter 8 and Fig 8.1.

13.7 Expediting

Project controls – the feedback loops of measurement and remedial action to control performance according to plan – are customarily split into cost, time, and quality functions. The time control system related to procurement is customarily referred to as 'expediting', although the term could be and sometimes is applied to any project activity (or the perceived lack thereof), including engineering.

Inevitably, any activities directed towards the expediting of vendor activities have an interface with other project activities which are related by physical sequence or information flow – expediting is a two-way activity. For instance, as regards the procurement of equipment, it is also necessary to expedite:

- delivery to the vendor of any outstanding design information;
- approval, by the project team, of vendor drawings before manufacture commences;
- concession requests (for deviation from the specification or from approved drawings) as manufacturing difficulties arise;
- queries on the interpretation of drawings; and
- third-party inspection and witnessing of tests.

It is just as important to expedite the flow of information from the project team to the vendor as it is to expedite the vendor's own activities.

Many quality authorities argue that expediting must be separated from inspection, in both line management and execution. No one person should share both responsibilities and be exposed to the ensuing conflict of objectives. Besides, the respective roles tend to require different skills and personalities – the quintessential inspector being a skilled and meticulous technician, while the expeditor may be perceived as a bully who will accept no delays on any account. Needless to say, this is an over-simplification which grossly undervalues the maturity of some of the professionals involved. While the ideals of role separation may be preserved with regard to more critical items, the dictates of economy make it inevitable that there is a level of criticality below which the roles of expediting and inspection are combined.

For the most part, observations made in the preceding section about the planning of inspection, may be considered interchangeable with those on expediting, provided that quality is substituted by time. Both functions require a plan as a basis of control, and a plan of execution based on the criteria of criticality and limiting returns. And both functions are worthless without remedial or preventive action based on the observations made, the facility for which must be incorporated into the agreement of purchase.

A standard purchase-order expediting report needs to reflect the status of:

- finalization of the purchase order and amendments;
- information flow between the vendor and the project organization as listed above;
- delivery to the vendor of 'free issue' parts;
- progress of the actual work being done by the vendor, and his plans;
- orders on sub-suppliers; and
- formal inspections, acceptances, and rejections.

The report should also list the *original* planned date for completion of each milestone, the current prediction, and the corrective action.

Chapter 14

Fluid Transport

In the following three chapters on materials transport systems, we will outline some of the basic concepts and their application to a process plant project. It is of course impossible to go into much depth in a few pages. We aim to establish an idea of the discipline scope to be incorporated within a well-engineered project, and provide some references by which a potential engineering all-rounder may develop his capabilities for use in situations where specialized knowledge is less important than rapid and inexpensive conclusion.

Here we will address contained fluid handling systems; the open-channel flow of fluids is included in Chapter 17, together with plant drainage. We will begin by addressing the transport of liquids in pipes. The transport of liquids in containers may be important outside plant limits, but it has little place in a continuous process plant and will not be discussed.

The essential elements of liquid-handling systems are vessels (which are referred to as tanks when open to the atmosphere, and pressure vessels when the liquid surface is under pressure or vacuum), pumps, piping systems, and valves.

14.1 A brief note on liquid-pumping systems design

For process plant design, it is convenient to calculate pumping performance in terms of head of liquid rather than by pressure. The head (h) is the height of the vertical column of liquid which its pressure (p) will sustain, and the two are related by

$$h = \frac{p}{\rho g}$$

where ρ is the fluid density and g is the acceleration due to gravity.[1]

Except for the influence of viscosity variation (which is not very important in many in-plant pumping systems), the head required to pump a given volumetric flowrate of liquid (Q) through a piping system is independent of the fluid pumped, and in particular its density, as is the head generated by a centrifugal pump. Hence for such systems the head/flow characteristics are practically independent of the composition of the fluid pumped.

The single-stage or two-stage centrifugal pump is usually the preferred pump type in process plants, when the performance range of such pumps is suitable. The advantages include high reliability and ease of maintenance, which arise out of simplicity of design and lack of internal rubbing parts, and flexible performance characteristic over varying flowrate. A good understanding of the features of this class of equipment is a sound investment for process plant engineers.[2]

Positive displacement pumps are used when the head required is relatively high and the flow is relatively low. Multi-stage and high-speed centrifugal pumps also compete for duties within this range, and are often preferred, but their ruggedness and simplicity are diminished by such features as inter-stage bushings and gearboxes. Positive displacement pumps have other characteristics which make them useful for special duties, for example their essentially constant flow characteristic under varying discharge head is useful in metering applications. It is usual that a certain pump type becomes the established choice for a given process duty. The single-stage centrifugal pump is the default type.

The centrifugal pump consists of an impeller, which transmits energy to the fluid (both by imparting velocity to the fluid and by the passage of the fluid through a centrifugal force field), and a stator or volute – the pump casing – in which kinetic energy is transformed into pressure energy or head. The head imparted (h) is proportional to the square of

---

[1] Any consistent units may be used; for example, in this case h is in metres, p is in Pascals (N/m^2), ρ is in kg/m^3, and g is in m/s^2.

[2] A classic reference is A. J. Stepanoff, *Centrifugal and Axial Flow Pumps: Theory, Design and Application*, 1992.

the impeller tip speed (v), which is in turn proportional to the product of impeller diameter (D) and rotational speed (n). Summarizing these basic relationships:

$$h = k_1 v^2 = k_1 \pi^2 n^2 D^2$$

where k_1 is a number which is approximately constant for a given flow pattern in the pump, or ratio of flowrate to tip speed for a given impeller shape. Most pump models can employ a range of impeller diameters, with variation of performance generally as indicated by the equation.

An important characteristic of the impeller is the angle of its vanes to the tangent. If the angle at outlet is 90°, that is, the vanes are radial, then the velocity imparted to the fluid by the impeller is substantially independent of the flowrate, and the pump's performance curve of head against flowrate is essentially horizontal, although curving downwards at flowrates above and below the design flowrate, due to inefficiencies. If the vanes are swept back, the fluid exit velocity has a tangential component which is proportional to the flowrate; as this component reduces the net fluid velocity, the head generated is correspondingly lower. Thus the performance curve is sloped downwards at increasing flow, at an angle which reflects the vane angle. It is also possible to build impellers with forward-sloped vanes to give a rising head–flow curve, but the performance is relatively unstable and inefficient, and such impellers are not used in practice. Radial-vaned (or nearly radial-vaned) pumps inherently develop higher heads than with back-sloping vanes (for a given tip speed), and are often employed on high-head duties.

The constant k_1 can be theoretically evaluated under ideal conditions by Euler's analysis, resulting in the conclusion that

$$k_1 = \frac{\phi}{g}$$

where ϕ, the head coefficient, is 0.5–1.0 depending on the vane angle and flow conditions. Losses in the actual case result in a lower coefficient. For a normal back-swept impeller, the value of ϕ at the best-efficiency flowrate could typically be in the region of 0.4–0.6. It is thus very easy to get a rough idea of the relationship between impeller tip speed and head under design operating conditions.

The power (P) drawn by the pump is given by

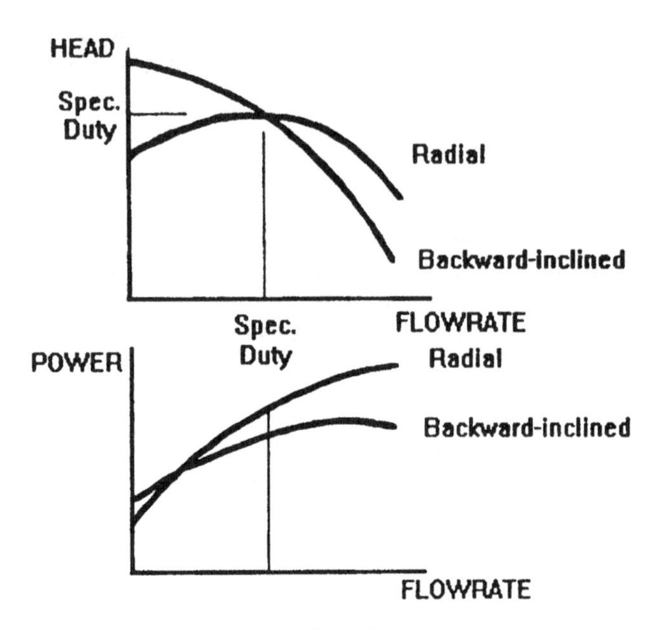

Fig. 14.1 Typical performance curves

$$P = \frac{Qp}{E} = \frac{Qh\rho g}{E} = \frac{hmg}{E}$$

where E is the efficiency of the pump, and m is the mass flowrate. The efficiency versus flowrate curve generally has a maximum at the pump design flowrate, and curves downwards at lesser and higher flowrates.

The power-versus-flowrate curve can be deduced from the head-versus-flowrate and efficiency-versus-flowrate curves for a given fluid density, using the equation above. It is found that the power curve tends to rise continuously as flowrate increases for a radially vaned impeller, while for a backward-sloped impeller the power rises less steeply (due to the head decrease) and may reach a maximum value and then decrease, which is described as a 'non-overloading' characteristic. The curve shapes are summarized in Fig. 14.1.

We will turn now to the piping system. Pipelines should theoretically be sized by a process of economic optimization, balancing the increased capital costs of greater pipe diameter against the increased pumping costs of reduced diameter and consequently higher head. Considerations of heat loss, abrasion, and pump cavitation may sometimes also affect the optimization process. In practice, it is usual to select as design basis a range of standard pipeline velocities, for example 1.0–2.0 m/s for pump discharge lines of increasing diameter, and 0.5–1.0 m/s for suction lines. The economics of such design criteria should be reviewed for the circumstances of individual major projects, especially where construction materials or power costs are unusually high or low, and where individual lines are very long or expensive. The design basis should be qualified in terms of whether the flowrate used is the normal or maximum value, and the implications taken into account when setting up the standards for the project. We will assume in the following that the design is based on maximum flowrate.

The dynamic (or frictional) head loss h_f may be calculated from

$$h_f = 2f \times \frac{V^2}{g} \times \frac{l}{d}$$

where f is the Fanning friction factor for the Reynolds number and the pipewall roughness, from standard hydraulic charts (for example from Perry's *Chemical Engineer's Handbook*). (Note: we are using the Fanning factor because we have quoted Perry as a reference manual. Be careful: Darcy's factor is also commonly used – this is four times the Fanning factor, so the formula must be adjusted accordingly). V is the mean fluid velocity in the pipe of length l and diameter d. For a given flowrate, V is inversely proportional to the square of the pipe diameter. Allowing for the decrease in friction factor as the pipe diameter increases, approximately

$$h_f = \frac{k_2}{d^5}$$

Given this exponential relationship between head loss and diameter, the advantages of rounding up to the next pipe size when in doubt should be obvious, as well as the severe penalty of undersizing the pipe.

The majority of process plant pipelines operate at moderate pressures, say less than 10 bar. Consequently, as can be verified by simple hoop-stress calculations, for steel pipes the pressure-induced stresses are fairly nominal in relation to the material strength, and it is not necessary to increase the thickness much as pipe diameters increase. Pipe wall thickness is selected to provide adequate corrosion allowance and mechanical strength and rigidity, particularly in bending. This is not the case for plastic pipes, where more care must be taken to avoid pressure-induced failure, and it becomes more important to understand system characteristics by which over-pressure may arise.

Pipe support configurations are an important facet of design. A small increase in diameter greatly increases the flexural rigidity of a pipe span, which is proportional to d^4, and therefore increases the permissible support spacing (the load of the fluid content increases only as d^2). Support considerations are rather different in the case of plastic pipes, which often need full-length support in the smaller sizes to avoid excessive deflection and bending stresses.

Process plant pipelines are invariably built according to a number of standard material specification systems drawn up for each plant. The material specification for each category of pipeline establishes both the materials of construction and the purchase specifications of the building blocks of pipe, fittings (flanges, bends, tees, etc.), and valves. There are usually also some specific design or test and inspection details, for example special bend radii, radiographic inspection, or hydro-test. Material specifications are compiled for different ranges of pressure and temperature, usually on the basis of the pressure–temperature rating of standard flanges, for example ASME Standard B16.5. There are also different specifications for different process fluids, requiring special materials or coatings to resist corrosion and possibly wear. The library of specifications for a plant or project is drawn up fundamentally by economic optimization. Too few specifications will result in unnecessary expenditure on lines which are overdesigned, because their duties fall into the less severe end of a wide duty range, while the use of too many specifications over-complicates construction and maintenance, and diminishes economies of scale. The pipeline sizes and material specifications are both indicated on P&I diagrams.

Having established the basics of pipeline diameter and piping specification, we can progress to the performance of the combined pipe/pump system. The head required is calculated from its static (h_s) and dynamic (h_d) components.

$$h = h_s + h_d$$

To calculate h, firstly the pipeline configuration is established, from suction vessel to discharge vessel or outlet fitting. Usually this is carried out in sketch form, based on the initial layout, because the information is required to order the pumps so that the final piping arrangement can be drawn up incorporating the actual pump details. There is usually a degree of iteration here, because the pump selection must ultimately be checked against the final piping design.

h_s is calculated from the relative liquid levels and pressures of the suction and discharge vessels. Usually there is a range of values, depending on the possible variation of level and pressure. h_d is calculated by adding to h_f an allowance for the pressure loss in fittings, valves, heat exchangers, etc. in terms of data available for these items (refer to Perry for standard coefficients for fittings), which may be expressed in equivalents of length of straight pipe or as a multiple of the velocity head $V^2/(2g)$.

To check the performance of a pump over a range of anticipated conditions, it is usually acceptably accurate to use the approximate relationship

$$h_d = k_3 Q^2$$

where k_3 is established from the dynamic head calculated for the flowrate for the defined normal duty.

The system head envelope for different static heads (and, if required, different control valve settings) can now be plotted against the flowrate,

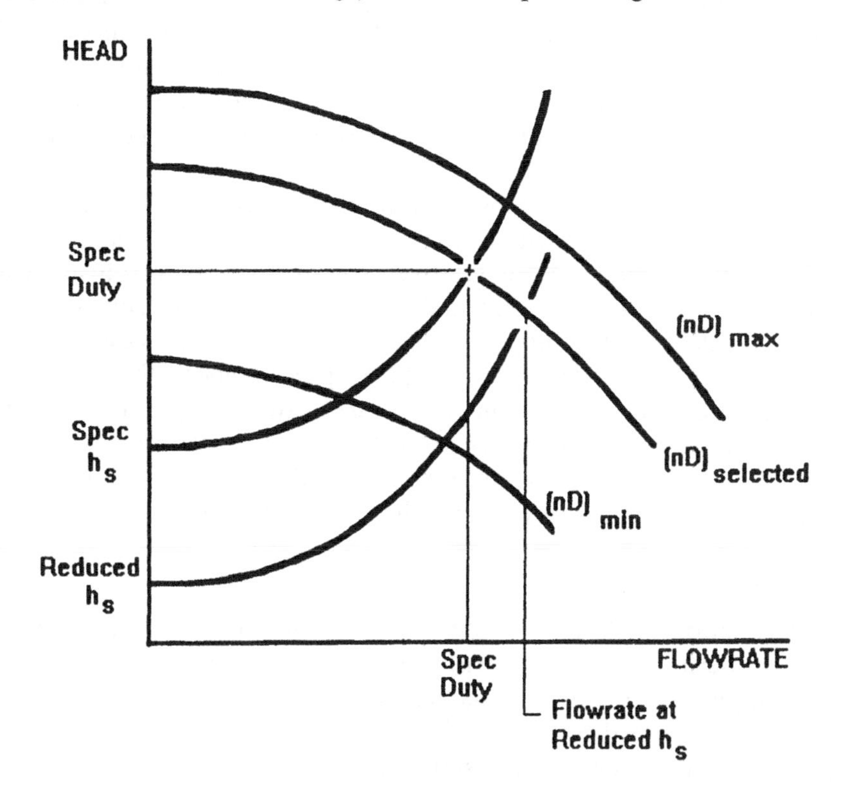

Fig. 14.2 Pump performance and system resistance for varying static head (h_s) and impeller tip speed (nD)

and the pump performance envelope for different impeller diameters and speeds can be superposed, Fig. 14.2. The intersection of the curves obviously gives the flowrate for a given pump and flow system, and gives the corresponding head and power draw. Utilizing the curves, it is possible to evaluate the pump suitability over various process conditions and the flow controls that may be required. Control, if necessary, may be achieved by using a control valve, or by pump speed variation by a device such as a variable frequency drive, which saves power but is usually more expensive to install.

The pump driver has to be rated for the maximum possible power draw. In the case of an electric motor (by far the most common choice), its protection will cause it to trip if overloaded, upsetting the process operation. Thus a pump with a non-overloading characteristic (and therefore downward sloping head curve) may be preferred for duties where the maximum flow is not easily limited, in order to lower the necessary drive motor rating.

It is also necessary to have an adequately downward-sloping curve to control load-sharing when pumps operate in parallel, and in order to maintain stability when pumping against a static head with very little frictional head. However, for most process duties a relatively flat curve is tolerable, with the advantage of relatively high head per pumping stage.

Positive-displacement pumps have an almost vertical head–flowrate curve; the decline in capacity at increased pressure results mainly from increased internal leakage, a relatively small quantity in an efficient pump. It is not, however, normal to present the performance of a positive-displacement pump on a head–flowrate basis; it is essentially a constant flowrate device (for constant speed), where the discharge pressure is determined by the discharge system only. Such pumps usually require a relief valve as protection against overpressure in the event of flow restriction (typically, a closed valve), and capacity is controlled by a bypass flow rather than by a series control valve.

Visualizing heads

The calculation of fluid flow and pump performance is usually carried out in modern design offices with a computer, and many programs of varied sophistication are available. However, reliance on computers may be accompanied by a lack of understanding of what is really going on, and with this lack of understanding, intuitive knowledge on system design may also be lost. It also becomes more possible to make an order-of-magnitude error, if computer input is incorrect!

For the uninitiated reader, it is worthwhile performing a few exercises in pump head calculation and pump performance to develop a feel for what is going on in a pumping circuit.

As a simple example, consider a pipe of nominal bore 250 mm, containing a fluid with mean flow velocity 2 m/s (which as we remarked is towards the upper end of commonly applied in-plant flow velocities for clear fluids). The velocity head

$$h_v = V^2 / 2g = 2^2 / 2 \times 9.81 \approx 0.20 \text{ metres}$$

and this is the minimum head needed to get the fluid moving out of the suction vessel.

To estimate the frictional head loss in the pipe h_f, note that the Fanning friction coefficient f, for typical non-viscous applications, is usually in the range 0.004–0.008, so if we take it as 0.006:

$$h_f = 2f \times (V^2/g) \times (L/d)$$
$$= 2 \times 0.006 \times (2^2/9.81) \times (L/0.25) \approx 0.02 \times L$$

This may be expressed as 2 per cent of the pipe length.

Now consider the pipe fittings, if they are clear bore items such as elbows or gate valves, one would surely expect the associated head loss to be a fraction of one velocity head per fitting. Little of the velocity energy is lost through such fittings, provided that a minimum of eddies and suchlike turbulence is created. The 'loss tables' for various types and dimension of fitting confirm this, refer to Perry's *Chemical Engineer's Handbook* or other tables for typical loss values as a ratio of velocity head. The tables also show significantly higher loss values for items such as tees and globe valves, up to a few times the value of the velocity head, as one would expect. However, for an 'average' in-plant case, one can expect the losses through the fittings and pipe-entry, assuming that there is no control valve, to be of the same order of magnitude as the frictional loss, i.e. another 2 per cent or so of pipe length.

Turning now to the pump, if this is a single stage centrifugal pump, we can derive the head from

$$h = \frac{\phi}{g} \times (\pi n D)^2$$

We expect ϕ to be in the range 0.3–0.7. If we take it as 0.5 and assume a pump of impeller diameter 300 mm rotating at 25 rev/s (4-pole speed in a 50 Hz system), we have:

$$h = 0.5/9.81 \times (\pi \times 25 \times 0.3)^2 \approx 28 \text{ metres}$$

If the static head is 20 metres, and the pipeline is 100 metres long, one would expect a total system head requirement in the order of 25 metres, so this should be well within the range of available centrifugal pumps, without resorting to 2-pole speed.

14.2 Gases

For the most part, the design of gas transport systems follows the same lines as for liquids, except for the phenomenon of compressibility and consequent density variation. System performance is presented on a head basis in the case of turbo-compressors or fans, which corresponds to centrifugal and axial flow pumps, in the same way as for pumps. There are applications, such as most flue gas or ventilation systems, where the density variation is slight, and the same equations and design systems can be used as for liquids, except that economical pipeline design velocities are usually in the range of 10–30 m/s, the higher values being for higher diameters and lower gas densities.

When the effects of density variation cannot be ignored, for example with pressure ratios (maximum to minimum within the system) of greater than 1.2:1, calculation of the head requires the assumption of a relationship between the pressure and density (or its inverse, specific volume ν) of the gas. The two most widely used expressions are the 'polytropic' relationship

$$p\nu^n = \text{constant}$$

and the 'adiabatic' relationship

$$p\nu^\gamma = \text{constant}$$

The polytropic exponent corresponds to the actual relationship for the particular compressor in question, and varies (for a given gas) depending on the efficiency of the compressor. The adiabatic relationship (in which γ is the specific heat ratio for the gas) is independent of the compressor performance and thus a little less accurate, but is useful in practice owing to simplicity.

The polytropic head is given by

$$h = \left(\frac{n}{n-1}\right)RT_1\left[\left(\frac{p_2}{p_1}\right)^{\frac{n-1}{n}} - 1\right]$$

where h is the head, R is the gas constant, T is absolute temperature, p is absolute pressure, and subscripts 1 and 2 denote suction and discharge. The adiabatic head may be calculated by substituting n by γ. h may be expressed conveniently in energy per unit mass, which corresponds to the equation above. If R is expressed in kJ/kg °C, T must be in K and h will be in kJ/kg.

h may be expressed in metres (as is usual for pumps) by dividing the J/kg by g

$$h(\text{metres}) = \frac{h(\text{J/kg})}{9.81(\text{metres/s}^2)} \approx 100h(\text{kJ/kg})$$

The power absorbed is given by the same expression as for a pump

$$P = \frac{h(\text{metres})mg}{E} = \frac{h(\text{J/kg})m}{E}$$

where P is the power, m is the mass flowrate, g is the acceleration due to gravity, and E is the efficiency.

A further useful relationship is

$$\frac{n-1}{n} = \frac{(\gamma - 1)}{\gamma E}$$

from which it is seen that n and γ are equal when the efficiency is unity, as one would expect. The head equation is in fact not very sensitive to the value of n. When calculating the head required for a typical process application, it usually suffices to take E as 0.7, at least for a first iteration, unless there is better knowledge of the efficiency of the intended compressor.

Example

For an air compressor, the operating conditions are

Inlet pressure = 1 bar absolute
Discharge pressure = 8 bar absolute
Inlet temperature = 20 °C = 293 K
Flowrate = 300 Nm^3/h ≈ 360 kg/h = 0.1 kg/s

Evaluate the polytropic head and power absorption, assuming that the polytropic efficiency is 70 per cent.

Solution: for air, $R = 0.29$ kJ/kg K and $\gamma = 1.4$, so

$$\left(\frac{n-1}{n}\right) = \left(\frac{(\gamma-1)}{\gamma E}\right) = \left(\frac{1.4-1}{1.4\times 0.7}\right) = 0.4$$

$$h = \frac{0.29\times 293}{0.4}\left[\left(\frac{8}{1}\right)^{0.4} - 1\right] = 276\ \mathrm{kJ/kg}\quad (\text{or } 27\,600\ \mathrm{m})$$

$$\text{Power, } P = \frac{276\times 0.1}{0.7} = 40\ \mathrm{kW}$$

This is not a practical case! Intercooling would normally be used. If we consider for instance a centrifugal plant-air compressor, three stages may be employed. If we ignore the pressure drop through the intercoolers, and assume that they can return the air temperature to 20 °C, the pressure ratio per stage would be $\sqrt[3]{8} = 2$. Reworking the expression above yields a head per stage of 68 kJ/kg and 204 kJ/kg in all. The absorbed power becomes 29 kW.

Although a head/stage of 68 kJ/kg can be attained by the geared high-speed units employed in packaged air-compressors, for normal process purposes a figure of around 30 kJ/kg (3000 m, or 10 000 ft) is about the maximum head per stage, and is recommended for preliminary purposes, for example when deciding whether to employ a single-stage centrifugal blower or a positive-displacement blower.

Euler's analysis and the resulting flow coefficient is essentially the same as for a pump, and on the same basis we could expect that for a stage polytropic head of 3000 m, we would be looking at an impeller tip speed of something like

$$v = \sqrt{\frac{gh}{0.5}} = \sqrt{\frac{9.81\times 3000}{0.5}} = 250\ \mathrm{m/s}$$

The analysis presented above is rather crude, and to be used for order-of-magnitude purposes. For gases which behave less like ideal gases, it is necessary to compute the head as a change of specific enthalpy and make allowance for compressibility, rather than use the formula above.

As in the case of pumps, centrifugal (and for lower heads with higher

Fig. 14.3 Centrifugal compressor installation. Note that the designer has chosen to locate the connecting piperack on a level above the compressor; this provides the facility to anchor the piping loads caused by thermal expansion, and also maximizes access to the compressor (while minimizing the elevation). Note also the bypass control valve for surge protection

flowrates, axial flow) turbo-compressors tend to be preferred within the range where they are effective. For higher heads or compression ratios, positive-displacement compressors are required. As for pumps, the virtues of certain types of compressor have come to be recognized for certain process duties, and experience must be considered when making a choice. For the most widely required application – the production of plant utility compressed air – fundamentally different types of compressor, including geared-up centrifugal machines, screw-type machines, and reciprocating machines, are able to compete. Each type has a preferred niche within its most effective flow–pressure envelope, or for other considerations such as oil-free air requirement.

A special word of caution is required in relation to the application of turbo-compressors. Although the same general considerations apply as for pump curves of head against pressure, the compressor curve almost invariably exhibits a reduction in head at low flows. In this region, flow is unstable and surging will be experienced, that is, periodic flow oscillations. These are liable to cause stress and overheating, which the compressor may not be able to withstand, and which are anyway not usually tolerable for process performance. It is essential to check that the minimum flow requirement of such machines does not overlap the range of process operation. It is possible to protect the machines by installing bypass systems coupled to minimum flow detectors – these often come as part of the package – but the protection systems may be relatively expensive, and should be taken into consideration at the time of purchase (not when the compressor has been reduced to a pile of scrap!).

In general, suitable experience should be brought to bear when ordering compressors and pumps, especially for expensive applications. Considerations should include the following.

- Suitability for the full range of process duties. This includes staying clear of compressor surge, as mentioned above, and also rotational speeds which are too close to critical values.[3]
- Potential start-up and shutdown problems.
- Efficiency and power cost, at normal duty and other required duties.
- Adequacy of shaft sealing arrangement and other rubbing seals where applicable.
- Facilities required and available for cooling.
- Capacity control.
- Adequacy of instrumentation and protective devices.
- General maintenance requirements and reliability in relation to process duty; possible need for an installed standby unit.
- Availability and cost of spare parts and service.
- Standardization, where possible, of components such as shaft couplings, bearings instrumentation, and seals. (Shaft seals are the most frequent maintenance item on a pump.)
- Above all, the track record of the application of similar machines to the process duty in question, and the track record of the machine vendor under consideration.

API (American Petroleum Institute) standards are commonly referenced for the purchase of compressors and pumps for oil refineries

[3] In the author's experience, failure by process technologists to specify the full range of operation is the most frequent cause of the selection of inappropriate equipment, especially for turbo-compressors. The duty specifications should be proactively cross-examined.

and demanding process duties. However, these standards are often too demanding for light-duty applications, such as pumping water or compressing air, and non-continuous duty. There are many national industrial standards applicable to 'chemical', water, and air or flue-gas duties. Engineers unfamiliar with these may be better advised to adopt simple functional specifications, coupled with the input of adequate and relevant experience, to evaluate what is offered. API standards are not at all applicable to most metallurgical applications, which centre on slurry pumping, and for which vendor standards often suffice.

14.3 Piping engineering and its management

Having given some consideration to the sizing of pipelines and specification of appropriate pipeline materials systems, we will return to the subject of piping engineering. While this at first seems to be a relatively unchallenging subject, especially for low-pressure and ambient-temperature applications, it is actually the most frequent cause of cost and time overrun, disappointed clients, and bankrupt contractors. The challenge set by piping is not so much technological – for the vast majority of lines, tried, tested, and well-documented design systems are employed – but rather the vast amount of detail. This work is routinely underestimated, with ensuing chaos as attempts are made to perform the work within schedule, both in the design office and on site. Piping engineering is above all a management task. The importance of planning in detail and working to plan, with attention to detail, cannot be over-emphasized.

The starting point for piping design, as for all disciplines, is the development of design criteria. The essence of these is the system of materials specifications, as previously discussed, but several other aspects must be addressed. Previous project design criteria are invaluable as a starting point; the following are a few of the more important contents.

- A basic standard for piping engineering, for instance ASME Standard B31.3 for process piping is commonly employed. This addresses subjects such as the calculation methods to be employed for design and stress analysis, the treatment of thermal expansion, and the permissible stresses for various materials and temperatures.
- Standards of pipework fabrication, welding, marking, erection, testing, painting, and cleaning, in complement to the design standard employed.

- Specification of the preferred or required means to allow for thermal expansion. Often expansion joints or bellows are considered impermissible on grounds of reliability, and the provision of inherently flexible piping design (incorporating sufficient bends) is specified.
- Any detailed requirements of valves needed to supplement the materials specification data, and similar supplementary data and standard specification cross-references for pipe and fitting materials.
- Standard and special dimensions and designs to be employed, for example for clearance under walkways, arrangement of expansion loops, installation of insulation, and pipe supports. Special process requirements must be considered in this context, for instance if there is a need for avoidance of pockets of stagnating process material, and consequent orientation of branch connections, etc., or piping that is rapidly demountable for cleaning.

The design of process pipework has to be co-ordinated with the construction methodology and organization. A large number of components, pipes, fittings, valves, etc. have to be ordered and stored. The fabrication and on-site construction of each line have to be controlled, sometimes with stringent quality specifications requiring the identification, approval, and traceability of each component and weld. The management of these tasks is most often handled by providing an individual isometric drawing for each line, which includes the full detailed material list and can readily be used as a control document for materials, work progress, and inspection. Constraint of time and control of detail usually dictate that the isometrics should be produced by the piping design office, in conjunction with the general arrangement drawings, rather than by the construction organization. This need not always be the case. Production of isometric drawings is often the most manhour-intensive project design activity, and one way to reduce the apparent cost of engineering is to leave the production of isometrics (or whatever rough sketches or lists will suffice) to the piping fabrication and erection contractor. This has the added benefit that it obviates claims and disputes between the designer and constructor over the consequences of errors in the isometrics.

Ducts – systems of pipes (usually made from thin rolled plate) operating at near-atmospheric conditions – are worthy of a special mention as being the cause of unnecessary problems. Some design offices regard these items as 'piping' and some do not, with the result that initially they may receive insufficient attention, being no-one's responsibility! However, the diameters may be large, requiring special attention at the plant layout stage, and resulting in horrible consequences if this is delayed owing to oversight. Large diameter ducts in hot applications usually require expansion joints, as expansion loops are too cumbersome, and these joints need special attention, both in their detailed design and with regard to duct support. Comparatively modest pressures or vacuums in large ducts can result in formidable loads being imposed on supports, because of the reaction caused by unbalanced pressures at uncompensated expansion joints.

The fundamental tool for the management of piping work is the pipeline (or line) list. This lists each pipeline and usually includes its number, process service and material specification (possibly included in the numbering system), origin and destination, diameter, and process conditions of pressure and temperature. Test pressure and requirement for insulation may also be shown, if not considered obvious from the design pressure and the material specification system, and there may be a requirement for special remarks, such as the need for pickling the line.

Problems sometimes encountered in the preparation of this list include the following.

- Omission of lines, usually as a result of omitting the lines from the P&I diagrams on which the list is based. In estimates and the initial stages of the project, by far the major cause of omission is incomplete process design work, such as systems whose control or pressure-relief requirements have not been fully worked out under all process conditions, and insufficient attention to utility requirements. For correct cost-estimating and work planning, it is imperative to make adequate allowance for this by painstaking review of the process design and by drawing on previous experience. In the later project stages, common problems are interfaces with proprietary equipment (say, cooling water manifolds), temporary lines needed for commissioning, and piping items around vessels ('vessel trim').
- Incorrect statement of pipeline design conditions as a result of lack of consistency between the process engineers' practice and the requirements of the piping design code.[4]
- Specification changes within pipelines. It is advisable to create a new line number if a specification change is necessary, and anyway the means of making the specification break and the consequences (for example to hydraulic testing) need careful attention.

[4] The piping engineer is strongly advised to check that the process engineers understand the piping code definitions and usage of these parameters.

Once an accurate line list is established, progress and cost are controlled by estimating and continuously updating the drawing production, the materials required/ordered/issued per line, the fabrication and erection, and the cost of each line, and continually summarizing the quantities from the list. This sounds simple enough, but requires a lot of effort and perseverance. It is in the failure to maintain and manage this continuing detailed effort that control is often lost. If this happens, it often comes about because the initial piping budget was based on bulk quantities without a breakdown per line, which does not fully exist until the last isometric is drawn. There can be no reliable control of progress, materials, and cost except on the basis of a comprehensive line list backed up by a budgeted take-off per line, updated as design progresses.

14.4 Some concluding comments on piping layout for pumps and compressors

The design of the piping to the suction of a pump is generally more critical than the discharge piping, and this is also the case for compressors drawing from a relatively low-pressure source. Pumps have to operate above the NPSH (net positive suction head) required, which is a measurable characteristic of the pump model for a given flowrate and pump speed. The NPSH available is the difference between the absolute pressure at pump suction and the vapour pressure of the liquid, expressed in terms of head. Effectively, it is the head available to accelerate the liquid into the pump impeller without causing vaporization, which is damaging to the pump performance and the pump itself. In the case of a tank having a 3 m liquid level above pump centreline, with cold water at sea level (vapour pressure negligibly small, atmospheric pressure = 10 m of water), the NPSH available is 3 m + 10 m = 13 m, minus the frictional head loss in the suction line. If for instance the NPSH required by the pump at maximum flowrate is 9 m, the suction line losses must not exceed 4 m, and it would be prudent not to approach this value.

Irrespective of considerations of NPSH, the pump suction line should direct the liquid smoothly into the pump suction without any bends or flow disturbances close to the pump, and without pockets where vapours or air can be trapped, eventually to be sucked into (and cause damage to) the pump. Very long suction lines should be avoided, irrespective of the apparent adequacy of NPSH as calculated above, because there are liable to be cyclical pressure waves between the pump and suction tank at start-up which can cause damage (or seizure, in the case of non-lubricating fluid such as kerosene). The same problem is found under normal running conditions for positive-displacement pumps without suction pulsation dampeners: the solution is to keep the suction line short.

In the case of compressors, the fundamental performance parameter is pressure ratio, which has a direct relationship to the head. A compressor of any type tends to be rated to achieve a certain pressure ratio for a given flowrate. A restriction in the suction system is more detrimental to performance than a restriction in the discharge. If a compressor is rated to compress a gas from 1 bar absolute to 10 bar absolute, it will be able to maintain a pressure ratio of 10 under varying suction pressure. The temperature ratio (often the limiting machine characteristic) corresponds to the pressure ratio.[5] Thus a pressure loss of 0.2 bar in the suction line will result in a pressure of only $(1.0 - 0.2) \times 10 = 8$ bar in the discharge for the same flow, or for the same temperature rise per stage.

As vapour bubbles tend to damage pumps, so do liquid inclusions tend to damage compressors: suction systems must avoid any possibility of liquid build-up or carry-over.

[5] The polytropic relationship is $T_2/T_1 = (p_2/p_1)^{(n-1/n)}$.

Chapter 15

Bulk Solids Transport

'Bulk solids transport' comprises not only several fields of specialized technical expertise but also a major industry whose manufacturers have developed countless proprietary designs for various applications. This is a very brief overview, focusing on the project engineering and layout interface.

The movement of bulk solids can initially be split into two categories: firstly the flow of solids under the influence of gravity (bulk solids flow), and secondly the transport of solids by conveyor or moving container. Usually the two transport modes are combined in series, for example with a chute feeding onto a conveyor. Solids are also transported in a fluid medium by slurry or pneumatic transportation, which is considered separately in the next chapter.

15.1 Bulk solids flow

Factors which influence the flow of bulk solids include:

- the composition and physical properties of the material, in particular its density;
- the size, size distribution, shape and surface condition of the material particles;
- the amount of moisture or other lubricating (or fluidizing) medium present; and
- the shape and surface properties of the interfacing stationary surfaces (for example the chute).

Solid materials dropped onto a heap form a pile with a characteristic *angle of repose* which gives an indication of fluidity. It is generally lower for round or granular particles. The effect of moisture varies, depending on the amount of moisture and the composition of the material. A relatively small amount of moisture can have a binding effect and restrict the flow of solids, especially fines, whereas a large amount of moisture invariably promotes fluidity and may ultimately destabilize a fines stockpile.

Basic considerations which govern the design of solids-flow devices (chutes, hoppers, silos, etc.) within a process plant are the following.

- The height of drop in relation to a given flowpath, or steepness in relation to an inclined plane path.
- The need to avoid blockage, which will arise if the drop-height or steepness is insufficient, or if flow is unacceptably restricted.
- The need to avoid the potentially adverse effects of material-on-material or material-on-fixed-surface interaction, promoted by excessive velocities or impacts. The effects include wear, structural damage to the containment, material particle size degradation, dust emission, vibration, and noise.

The primary objective of design is normally to produce a solids flowpath which minimizes the drop-height (and therefore also the cost of the associated structure, and solids-interaction effects), while maintaining adequate and reliable flow conditions required for process operation and control. There is always some degree of wear at the flow boundary, and minimizing the wear and repair costs (which usually include consequential plant shutdown costs and loss of operating income) is also an important consideration. Wear of stationary surfaces is minimized by employing dead-boxes, in which the flowing material is retained (thus substituting material-on-material interaction for friction against the retaining surface or chute), and by employing replaceable abrasion-resistant chute and bin liners. The choice of liners ranges from hard metals to ceramic tiles, rubber, and plastics; chutes consisting only of a flexible rubber-type material such as a tube can also be very effective in the appropriate application. The composition and surface properties of the stationary surface or liner also have a major influence on surface friction and hence the required drop-height.

For most plants, the consequences of chute blockage are far more damaging than wear and other negative effects of excessive drop-height. Besides, once a plant has been built incorporating inadequate drop-heights, it is usually very difficult and expensive to increase them, whereas it is comparatively cheap and quick to change the chute configuration to slow down the material if the drop-height is found in operation to be excessive. It pays to be conservative by designing drop-heights at the top end of any range of uncertainty.

In any enclosed flowpath, chute, or silo the width of the flowpath has to be adequate, compared with the material lump size, to avoid blockage by arch formation – as a rule of thumb, a minimum of three times the maximum lump size.

15.2 Conveyors

Devices which move bulk solids may be roughly classified into two groups: those which transport material in discrete quantities by independently moveable containers, and those which convey a continuous stream of material. In general the first class is appropriate for lower capacities and higher material lump sizes, whereas the second class is preferred, wherever reasonably possible, for process plant applications because of compatibility with a continuous process.

We will therefore not dwell on containerized transport devices. The equipment types include wheelbarrows, skips, ladles, railtrucks, roadtrucks, overhead cableways, etc., and their usage is more often at the feed or product end of the process plant, or for collecting spillage.

The most widely preferred device for continuous material conveying is the troughed belt conveyor, its advantages being reliability, ease of maintenance, high energy efficiency, and, except for very short distances (say, less than 15 m), relative economy. There is a correspondingly voluminous quantity of literature available on its design and application. The Conveyor Equipment Manufacturers' Association of USA (CEMA) guides are recommended for general reference, being generally available, widely used, and well presented.

The CEMA guides can readily be used to establish the design parameters of most in-plant conveyors, and conveyor belt suppliers are usually happy to check and comment on the calculations, as a service. Longer or higher-power-draw conveyors, for example over 1 km or 250 kW, may require more specialist design attention because of dynamic effects relating to the elasticity of the belt.

The basic methodology for conveyor design follows.

1. Select the belt width and load cross-section formed by the troughing idlers. This choice is based on the characteristics of the material conveyed, the required maximum capacity, and the belt speed.
2. Calculate the power absorbed, this being the sum of power to elevate

the load, and various frictional losses, including hysteresis losses within the belt.

3. Calculate the belt tensions. The tension difference $T_{\text{effective}}$ required to move the belt is found by dividing the power absorbed by the belt speed. The maximum permissible ratio of tensions around the drive pulley is given by the frictional characteristics of the belt–pulley interface and the angle of wrap; it is typically around 2.5:1 for a single drive pulley. Considering the overall geometry and static calculation, the minimum slack-side tension to be induced by counterweight (or screw take-up on a short conveyor) and the tensions at all positions of the belt may be calculated.
4. Select a suitable belt carcass in accordance with the maximum belt tension, with rubber (or other elastomer) covering according to service requirements.

A few important points relating to the application (rather than the detail design) of belt conveyors will be addressed here. (Refer to Fig. 15.1.)

1. The design of the feed and discharge interfaces of a conveyor are critical to its performance. The best feed chute design will project the material onto the conveyor in a symmetrical stream at the velocity of the belt in the strict vector sense (magnitude and direction), minimizing wear, spillage, and dust. A poorly designed chute may load the belt asymmetrically (maybe more so at part load), resulting in material spillage and belt-rim wear along the length of the conveyor as the belt tracks off centre. It may cause excessive belt and chute liner wear by impact and abrasion, and spillage and dust

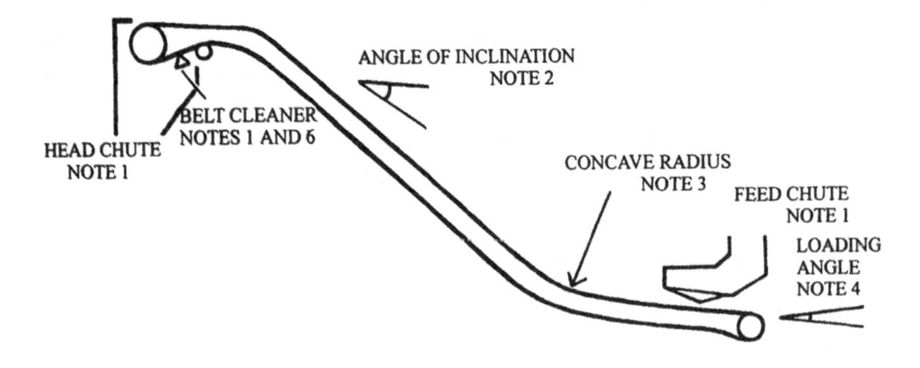

Fig. 15.1 Some important aspects relating to belt conveyors

due to turbulence and inefficient containment. It may block because of inadequate flowpath width or insufficient drop-height (especially in conjunction with change in flow direction).

The discharge chute design is less critical to the conveyor because the material flow on the belt is no longer influenced; the main consideration is to the next material destination, especially if it is another belt conveyor. However, there is a problem of containment, particularly for sticky materials and high belt speeds. One of the most problematic features of a belt conveyor is the belt cleaner, usually a scraping device. If this is inefficient and the material is even slightly sticky, tons of material per hour can go past the cleaner, and build up on the return idlers or fall below the conveyor. The chute must contain both the conveyor discharge and the 'dribble' falling off the belt cleaner, which is usually situated between the head pulley and the snub pulley (sometimes there are two cleaners in series, the first usually directly on the head pulley). The chute must also allow adequate access for maintaining the belt cleaner, which needs frequent attention.

A chute application requiring special attention is one between two belt conveyors in series and at an angle to each other. To ensure hat material is centrally, symmetrically loaded onto the second conveyor at all material flowrates, the reliable solution is to direct the first conveyor's discharge into an even, vertical stream whose centreline intersects the following conveyor's belt line, and then to present that stream to the following conveyor via a symmetric chute. This can require several metres of drop-height, depending on the acuteness of the angle between the conveyors, but any compromise may lead to uneven loading of the second conveyor and spillage.

It will be noted that although increase of belt speed proportionately increases the capacity, or decreases the width of a conveyor with the same capacity, high belt speeds are disadvantageous to the feed and discharge chute design and performance. In general, therefore, belt speeds above 2 m/s are seldom considered acceptable for the relatively short conveyors within plant confines.

2. The maximum permissible angle of belt inclination to avoid spillage is about 18°, less for many materials; tables for most applications are available in the CEMA guides.
3. Within a plant, belt conveyors invariably travel in a straight line as projected onto the horizontal plane. It is possible to introduce horizontal curves to the belt line by tilting the idlers away from the centre of curvature to oppose the belt tension, but the comparatively

large radii of curvature which are required are seldom appropriate within a plant. Curves in the belt line are frequently introduced in the vertical plane. Equilibrium of forces on the belt shows that a convex curve – such as from upwards-inclined to flat – can be built at a comparatively small radius, say 20 m; the force to oppose the belt tension is transmitted through the idlers, the spacing of which may be reduced to prevent any overload. A concave curve – such as from flat to upwards-inclined – requires a substantially greater radius, say 300 m, to prevent the belt from being lifted off the idlers. The consequences of this may include loss of alignment and spillage of material, a situation which is at its worst on start-up, with increased belt tension during acceleration, and when only the feed end of the belt is loaded with material. The minimum radius R to avoid belt lift is given by $R = T/mg$, where T is belt tension and mg is the belt weight per unit length.

4. The most expensive wearing item of the conveyor is the belt, which is subject to the most wear at the point of feed as the material is accelerated and stabilized. Wear and instability are promoted if the conveyor is fed where the belt is inclined, especially in the case of large, round rocks which may be induced to roll backwards. Particularly in the case of higher belt speeds, large rounded lumps, abrasive materials, and high material-impact energy at the feed point, it may be necessary to load at minimal if any inclination to the horizontal. When a belt conveyor's duty is to elevate the material with minimum horizontal displacement (a frequent requirement), deciding on the maximum acceptable inclination at the feed point requires careful consideration and compromise. (There is some guidance given in CEMA publications.) If in doubt, allow no inclination at all.
5. Given the angle of inclination (if any) at the loading point, the minimum concave radius, and the maximum inclination, it is now possible to set out the minimum length of conveyor for a required elevating height, for layout purposes.
6. Not all materials are suitable for transport on belt conveyors.
   - Wet materials with too-high fluids content may be difficult or impossible to handle owing to fluidization; the same sometimes applies to materials containing a lot of dry fines.
   - Sticky materials may be impossible to clean off the belt properly; installation of a belt washing station behind the head pulley is sometimes an effective though expensive solution, but raises additional complications of maintenance and disposal of

the washed product. Sticky materials may also be impossible to handle in chutes and bins.
- Big, round lumps tend to rotate and fall off inclined belts, causing a major safety hazard.
- Materials that are too hot will damage rubber belts, although by using neoprene or similar belt material, the permissible operating temperatures can be extended up to the range of 150 °C or so.

The issues listed above are some of the important considerations when laying out a plant incorporating belt conveyors, or in deciding whether a belt conveyor or system of conveyors is the best solution for a given application of bulk solids transport. The following alternative devices are often worthy of consideration.

- Belt conveyors having pockets or flexible slats (and possibly side walls), for increased angle of elevation without spillage. The disadvantages include more expensive and non-standard belting, and more difficulty in cleaning the belt. The higher-capacity, higher-elevation-angle arrangements, with flexible side walls, are *much* more expensive per metre than a standard troughed belt conveyor.
- Pipe-belt conveyors, in which the belt is wrapped completely around the material by rings of idler rolls (away from the feed and discharge points). Advantages include allowing the adoption of significantly steeper inclines than a troughed belt, and the ability to incorporate comparatively small-radius horizontal curves, thereby eliminating series conveyors and the associated transfer points. Disadvantages include more intensive maintenance requirements and the use of non-standard belts.
- Bucket elevators, which are a very economical option where straight elevation without horizontal displacement is required; they completely enclose the conveyed material and therefore eliminate dust emissions. They are relatively high-maintenance items, and are not suitable for use with very abrasive materials or large lumps.
- Chain conveyors, which have similar advantages and disadvantages to bucket elevators, except that horizontal translation and a path curved in the vertical plane are possible.
- Vibrating conveyors, which obviously have limitations on upward inclination, but can also fully enclose material. A succession of such conveyors is sometimes an economic choice to convey dry fines over moderate distances.
- Air slides, for fines, in which a comparatively low slope-angle is used in conjunction with fluidization by air.

There are several other types of conveyor, many of which have been found to satisfy a particular niche, and are often incorporated in a particular process technology package. There is one other type, the screw conveyor, which requires special mention because of widespread and traditional usage. Advantages of screw conveyors include:

- relative simplicity of design, manufacture, and maintenance;
- ease of dust-tight enclosure;
- facility to integrally include a feed extraction facility at the back end of the conveying section;
- possibility of providing multiple feed and discharge points on one conveyor; and
- The availability of ample information on many types of application, and the corresponding recommended design variations.

Features whose design may be varied to suit the application include the flight construction, percentage fill, speed, and design of intermediate hanger bearings (whose need may also be eliminated by appropriate shaft diameter and length).

There is a CEMA guide for screw conveyors, and much other useful literature, especially from manufacturers. A cautionary note is appropriate: whereas screw conveyors have been found to be the preferred choice for many applications, their choice has also been the downfall of several plants. Typical problems include the following.

- Application on an unsuitable material. For example: too sticky, or building up on the flights and shaft rather than flowing over them; too abrasive; or too lumpy, jamming between flights and trough.
- Failure to adequately de-rate the capacity and increase the power for upward inclination. The basic flow mechanism within a screw conveyor is that of material sliding down an advancing inclined plane; therefore it can be appreciated that the capacity decreases asymptotically as the inclination approaches the helix angle. (It is possible to improve the performance at upward inclination, including vertically up, by designing the conveyor as a fully flooded Archimedean screw, but the design basis is different and the power draw higher).
- Use of unsuitable intermediate (hanger) bearings in an abrasive application.
- Underestimation of the power requirement of the feeder section; this is addressed below.

15.3 Feeders

Feeders interface to the gravity flow of hoppers and silos, and promote and control the passage of materials onto conveyors and containers or into process equipment. There are very many types which have found application for different materials, material lump sizes, capacities, control requirements, and hopper or silo configuration. The most popular types include vibrating feeders, belt feeders, screw feeders, apron feeders, and rotary or star feeders. It is usually of critical importance to ensure that the type of feeder selected has a proven history of suitability for the application.

We will not consider the merits and details of various feeder designs: we will rather address the nature of the interface between the feeder and the bin of material above it, which impacts on the selection and design of the feeder, and is too frequently the cause of design error.

The feeder has to function with a volume of material in the bin above it. Obviously, if there is not always a positive volume of material it cannot produce a controlled output, and it is therefore not a feeder. Usually, the bin has a converging flowpath, say a cone, where it joins onto the feeder. The geometry, construction, and surface roughness of this flowpath affects the rate at which material can flow into the feeder, the vertical load transmitted onto the feeder, and the power required to drive the feeder. In the converging flowpath, frictional forces oppose the oncoming load of material, relieving the pressure applied onto the feeder.

The bin, or equally the flowpath down into the extraction point below a stockpile, must be designed to provide the required amount of 'live' (that is, accessible) stored material, without blockage. It is obviously unacceptable, and yet not unknown, for large silos or stockpile systems involving substantial investment to be useless due to susceptibility to blocking. Blockage is invariably associated with the design of the bin bottom and outlet. Obviously the outlet has to be large enough in relation to the material lump size, for example at least three times the greatest linear dimension. Equally, the slope towards the outlet must be sufficient to induce flow, although this may be promoted by the use of low-friction liners, vibrators,[1] fluidizing air (for fine materials), or airblast devices. A high slope-angle, say 70° or greater, may ensure the flow of all but the stickiest materials, but has expensive consequences to the cost of the retaining structure per unit volume of stored material: it

[1] Caution: in an unsuitable application (generally, with fines), vibrators may cause consolidation of the material, leading to blockage.

pushes up the height. An angle which is too low will result at least in dead material at the bottom of the bin, in a funnel-flow pattern which restricts the maximum withdrawal rate; it may also lead to the formation of a stable 'rathole' above the outlet, effectively blocking it.

Several practising bulk solids flow consultants have evolved systems to calculate appropriate bin and flowpath geometry, and selection of liner material, based on the measured properties of samples of the material handled (in particular the angles of internal and external friction). Such rationally designed systems also embrace the feeder interface, and generate design parameters which permit the confident selection of the right feeder. The accuracy of the calculations performed is obviously no better than the degree to which the samples tested are representative of eventual service conditions. The possibilities of variable material and moisture conditions, material consolidation over time, and service deterioration of liners must be considered.

In the case of major storage facilities, it is generally agreed that the capital expenditure involved warrants the cost of substantial testwork and consulting fees. For smaller installations, for example, hoppers containing say 20 m^3 of stored material, it is common practice for the experienced engineer to design the bin outlet geometry on rules of thumb related to previous experience with similar material. For instance, the following may be considered appropriate as a design basis *when a similar reference application exists to verify it.*

1. Minimum bin opening (width of slot) = the greater of 3 × maximum lump size (3 × 50 = 150 mm in this case) or the minimum successfully proven previously, which is 250 mm in this case.
2. Minimum valley angle in convergent section = 60°. Note: the valley angle is the true angle to the vertical of the line of intersection between two adjacent sides.
3. Assume that the downward load onto the feeder is equivalent to the weight of a volume of material equal to the area of the bin opening × the width of the feeder inlet slot (250 mm). This implies that the effective height of material equals the slot width: although there may be, say, 6 m of material above the opening, the load has been greatly reduced by the converging flowpath. This is a conservative value – the actual load is usually lower.
4. Assume that the maximum force to shear the material at the feeder interface equals the load due to material weight as calculated in point 3. (In other words, the equivalent coefficient of internal friction is 1, which is a conservative value that may be required at start-up; the actual running value should be lower.) This is the effective belt

tension in the case of a belt feeder, provided that it is correctly designed. In particular, the feed chute onto the belt must incorporate no converging material flowpaths; on the contrary, it must be relieved in the direction of flow by a few degrees, both in its width and in the elevation of the lower rim (loading shoe) relative to the belt. This permits material to be drawn down over the full length of the slot.

In the case of a screw-feeder, there is clearly a very complex variation of material pressure both over the height of the screw and in the space between the flights. It is possible to go to great lengths to compute the required torque, based on assumed variations of pressure – the following simplified method is probably as good as any. Compute the material pressure at the interface as outlined above, and apply this pressure over the total forward-directed area of screw flight in the feed area, to arrive at the thrust load. Apply a coefficlent of 0.3 to obtain a tangential force on the flights (allowing for both friction and the component of direct load). To calculate the torque required, the tangential force can be assumed to act at say 0.8 of the flight outside radius, depending on the shaft diameter. If the shaft diameter is large compared with the flight diameter, say over a half, then the circumferential frictional force and torque acting on the upper half of the shaft should be added. Note that the feeder section of the screw must also be designed for an increasing flow area along its length, by increasing the flight pitch along the length or by tapering the shaft diameter, or both.

It should be noted that filling the bin from near-empty while not extracting feed can result in abnormally high starting loads. These may be reduced by providing feeder supports which are relatively flexible, thereby inducing some movement of material relative to the bin as it is filled.

As a final note on feeders, in line with the comments made above: when feeders are purchased as an equipment item of proprietary design, the purchase specification should clearly include full details of the design of the bin above the feeder, as well as the properties of the bulk material.

15.4 Safety and environmental health

In conclusion to this section, the project engineer is reminded that the handling of solid materials can present major hazards to process plant operators. Reference should be made to BS5667, the equivalent of ISO1819, *Continuous Mechanical Handling Equipment – Safety Requirements*, which can be used as a reference specification when designing or purchasing a conveying system. BS5667 contains both design and operational requirements, and at least Part 1 should be thoroughly applied before authorizing a design for construction. An appendix to the standard lists the statutory regulations which are in force in various countries, and to which the designer must obviously adhere in that particular location.

The following are some of the aspects that need review and attention by the prudent designer.

- Exposed moving parts are hazardous, and there are always plenty of them around materials handling facilities. Belt conveyors, with the associated risk of pinching between the belt, rolling parts, and stationary parts, have always been potential killers and probably always will be, given the impracticality of complete enclosure. Guarding is necessary for all potential 'pinch points' (satisfying the parameters of BS5667 and national regulations). Pullwire stops are required for the length of the conveyor. Note that the guarding of conveyors can consume a significant number of design hours, and standard designs should be used wherever possible. The guards must be suitable for routine maintenance access to the conveyor for cleaning, adjustment, and greasing.
- In the case of accessible moving equipment parts that are started remotely, automatic audible alarms are needed before start-up.
- Pneumatic actuators moving accessible parts require special attention – they can stick and then jump.
- Consider dangers due to rocks falling off overloaded bins or open belts, and the containment of overflows in general.
- Provide safe access and a safe place to stand at potential blockage clearance and spillage removal points, and for all routine maintenance requirements such as belt cleaner adjustment.
- Dust is a hazard; it tends to be generated by dry materials at all open transfer points, which must usually be safeguarded either by the provision of dust extraction systems, or by wetting the material. Combustible dust (coke, feed-grain) may be an explosion hazard.
- Any high stacking of material requires care, owing to high loads and high potential energy. Vertical loading can cause foundation settlement, horizontal loading can cause collapse of abutting structures. Fine materials stacked high can generate a lethal mudrush when fluidized by water.

Chapter 16

Slurries and Two-Phase Transport

16.1 Slurry transport

Transportation of solids as a slurry in a carrying medium is widely practised in the minerals production industry, especially for hydro-metallurgical plant, where whole proses take place with fine solids suspended in water. It is not an economl transportation method when the water is simply added as a carrying medium that has to be separated from the solids at their destination, except for very long pipelines or for carrying solids which are in the first place too moist for handling by mechanical conveyors. Dry transport of solids, having appropriately fine grain size in relation to density, is routinely accomplished with air as the carrier in pneumatic transport systems. Separating the air from the dust at the destination can also be a relatively expensive complication, and pneumatic transport systems are therefore most competitive when the air is required as part of the process, as in pulverized coal burners fed by pneumatic transport systems, and in applications where air classification of the solids is required.

Systems for pumping slurries around a process plant are similar to systems for pumping liquids, in regard to the calculation of performance on a basis of head. The slurry can be regarded as a fluid having a density computed from the ratio of solids to liquid (usually water) and the respective densities of the two phases. If the slurry composition is stated on a mass basis, the slurry density is calculated by calculating the volumes of each component of mass, and dividing the sum of the masses by the sum of the volumes.

When pumping a slurry, the head generated by a centrifugal pump and the head corresponding to system resistance, are substantially unchanged from the values for water or any other low-viscosity liquid.

There is an increase in the apparent viscosity when there is a significant concentration of very fine solids in suspension, but for many applications the effect is negligible. Usually for in-plant applications, most of the system resistance is required to overcome static, rather than frictional, head.

System performance and power requirement are therefore normally graphically presented and calculated in the same way as for clear liquids. Subject to some minor corrections, a centrifugal pump will pump the same volumetric flowrate of slurry as for water through a given flow system. The pump discharge pressure and its power draw will be increased in proportion to the slurry specific gravity, as for any clear fluid of increased density.

One fundamental difference between the design of liquid-pumping systems and those for slurries is the need in the latter case to keep solids in suspension. Failure to do so leads to blockage and plant shutdown. There is a critical pipeline velocity for each slurry application, depending on the solids characteristics (in particular the density, size, and shape of the particles), on the liquid properties (we will assume it is water in all the following), on the solids/water ratio, and on the pipe diameter. Below the critical velocity the solids are liable to settle out, with an intermediate flow pattern known as 'saltation' at velocities just below critical. If settlement occurs, quite often it is difficult to re-establish flow, and there is an 'ageing' process which makes matters worse. The solids may simply have to be dug out, by dismantling the pipeline and/or using high-pressure flushing pumps. Onset of settlement is not only promoted by low velocity – turbulence caused by sharp bends, too many bends, other flow obstructions can be equally damaging.

In fact, particular slurries tend to have particular characteristics; these are not easily predicted theoretically, and only come to be known in practice. Some slurries are relatively easy to handle and, provided that the critical velocity is maintained, are unlikely to settle even with a convoluted flowpath, while other slurries will settle and block the line if only a few long-radius bends are present.

Another important factor is abrasion of the pipeline; corresponding fracture of the solid particles may also be an issue, when particle sizes need to be maintained. Abrasion obviously gets worse at higher velocities, in fact exponentially, so the slurry transport system designer cannot be too liberal in using higher velocities to avoid settlement, even if the necessary higher pump heads and power consumption are tolerable. Abrasion may be reduced by using appropriate pipe liners. For example, rubber is a very common and effective choice: the solid particles (up to a certain size, shape, and velocity) seem to bounce off it. However, larger and sharper particles may quickly tear and destroy the rubber. Quartz and metallurgical ores tend to be highly abrasive, while many chemical crystals are relatively unabrasive and can be conveyed at velocities where no liner is required.

Abrasion is usually a major consideration in pump selection. Except for the mild chemical-type duties already referred to, slurry pumps are of special construction, essentially to resist abrasion to the pump and its shaft seals but often also to permit fast maintenance and liner or component replacement for very abrasive duties. Generally either replaceable rubber lining or specially hard abrasion-resistant alloys are used; shaft seals are of special design, and are usually continually flushed with clean water. Impeller tip velocities are kept within proven limits for the application. Power transmission is often through V-belts rather than direct-drive couplings, to permit exactly the required speed to be obtained (rather than synchronous speeds) and to permit speed optimization in service (varying the impeller diameter is not so convenient).

There is a large volume of literature available on slurry system design, including formulae for critical velocity (the most popular is the Durand equation; refer to Perry, for details see p. 157) and for more precise flow resistance determination. These have their limitations however: design is best when based on experience of the duty. Much information can be obtained from the major slurry pump manufacturers, including Envirotech (Ash Pumps) and Warman. For critical applications, for example long tailings lines, it is wise to arrange for tests to be carried out on samples of the actual slurry, ultimately in closed circuit pumping loops, with interpretation and system design by an experienced consultant.

We stated above that for most applications the slurry behaves as a fluid of modified density, and that viscosity effects can usually be neglected. However, there is a limit to the permissible solids concentration. Relatively coarse solids can simply not be carried in suspension above a certain solids concentration for the particular slurry, say 40 per cent by mass. In the case of very fine solids, a 'thixotropic' mixture is reached at higher solids concentration, in which the solid particles stay in suspension but the slurry viscosity is markedly increased. In fact the viscosity no longer exhibits a Newtonian relationship of direct proportionality between shear stress and velocity gradient. This is an area to keep away from in normal plant design, but for special applications like mine backfill plants, when high solids concentration is critical, it may be desirable to commission a special design based on materials tests. It is also possible for special applications to employ a thixotropic (or at least enhanced-viscosity) carrying medium, obtained by high fines concentration, to carry coarse particles at relatively low velocities without settlement.

For most applications, in-plant slurry piping systems can be designed without any specialist technology by:

- ascertaining existing practice for the slurry composition in question;
- establishing the maximum value of workable slurry concentration (say 30 per cent solids on a mass basis);
- establishing the minimum allowable velocity (usually in excess of 2 m/s); and
- calculating performance as for water with increased density.

It is wise to allow for pump head and efficiency deterioration, relative to its water performance, of a few per cent, based on the pump supplier's data and according to the slurry properties. Indeed, the suppliers of specialist slurry pumps will usually assist in the overall system design. It is also wise to allow a margin of 15–25 per cent of maximum calculated absorbed power when sizing the driver, to provide flexibility for operation at increased densities and pump speeds.

16.2 Piping design

Piping design for in-plant slurry transportation is different from plain liquid piping in two important respects: the materials and components of construction, and the routing. Rubber-lined pipe is very popular, and requires that the pipes be assembled in flanged lengths of 6 m or so, with bends made as separate flanged items. High-density polyethylene is also popular; it is available in various grades, some marketed as ultra-high-molecular-weight polyethylene etc., with various claims made of superior abrasion resistance. Polyethylene and plastics generally are much weaker than steel and need more closely spaced and expensive supporting arrangements, and need to be supported continuously in smaller diameters. In less abrasive applications it is often more economic to install unlined mild steel pipe, and accept its replacement from time to time. Such practice should particularly be considered in conjunction with standby pumps installed with their individual, separate standby pipeline; keeping standby pumps with shared lines from blocking at the isolation valve is often problematical. The second line also confers other advantages of operational flexibility.

Valves for slurry duties are of special design. Control valves and check valves are not normally used, although some special valves (such as pinch valves) are available. Isolation valves have to be designed to avoid build-up of solids. Knife gate valves (in which the gate is designed to cut through solids), rubber diaphragm valves, ball valves, and plug valves are commonly employed; it is necessary for the engineer to verify the valve's application by relevant experience, just as when purchasing process machinery.

If a slurry stream flowrate has to be controlled, an attractive method is to use a variable speed pump. If the flow has to be divided the most usual system is to employ an elevated tank for the purpose, and control the divided flows with variable overflow weirs.

Returning to the subject of pipeline routing, we have observed that some slurries are more forgiving than others when subjected to a convoluted flowpath. It is essential to have a good understanding of these characteristics when laying out the pipelines. If in any doubt, assume that it is necessary to restrict the number of bends to a minimum, and keep their spacing far apart. Bends are subject to extra abrasion, apart from causing plugging of the line. In any event, long-radius bends should be used, and 45° entries should be used where flows converge.

It may not be easy to comply with the need for elimination of bends without creating serious access obstructions within the plant. When laying out a plant employing slurry transportation it is advisable to include the routing of the slurry lines, and certainly the more important lines, as part of the conceptual design. Verifying that these lines are as short and straight as possible should be a prerequisite to acceptance of the overall layout for detailed design. These lines may in practice be the most significant aspect of plant reliability.

Flushing facilities, for use when closing lines down and for blockage removal, are required for all slurry systems. Stagnant areas and unnecessary turbulence must be avoided. Pump suction lines require special care, in order to be kept as short and straight as possible and without excessive diameter.

16.3 Tanks and agitation

For systems handling slurries, special consideration has to be given to tank design to maintain the slurry in suspension. Usually an agitator is employed for this purpose. Quite often the tank is not just a holding vessel for materials handling purposes, it is part of the process, in which chemical action or crystallization takes place. There are several designs for the various applications, the most common being a round tank with diameter equal to liquid height, a central agitator consisting of an impeller pumping the fluid up or down (usually up), and four or more vertical baffles at the periphery to oppose fluid rotation. Another common design is the draft-tube type, in which there is a central vertical tube in which the impeller is housed; the liquid circulates up through the tube and back down the tank outside the tube.

Agitators are rated in terms of the process, as to whether heavy shearing is required for mixing or solid material attrition, or whether only maintenance of suspension is required. Scores of articles have been published on agitator design but there is no consensus on design method, at least as regards the sizing and hydrodynamic design. The engineer seeking bids on a basis of duty specification often receives, from experienced vendors, a wide range of basic designs (power input, rotational speed, pumping flowrate, impeller configuration, etc.) for the same duty. The amount of variation in designs offered is matched all too often by the number of unsatisfactory applications. Care is required in selection; this is an area where commercially competitive buying is a source of potential problems, in truth because it is not usually easy to be sure of the amount of agitation needed, and because the competition drives vendors to remove margins of safety.

Where possible it is highly desirable to develop basic design parameters from similar existing and performing applications, and use these as specified minimum design parameters (or at least as evaluations of the adequacy of equipment offered by vendors), together with the mechanical features offered. The most significant performance aspects are:

- the power input per unit tank volume (or mass of slurry);
- the pumping rate (on a basis of fluid velocity); and
- the ratio of impeller diameter to tank diameter.

The impeller design, dimensions, and speed should be compared for consistency with the claimed pumping rate.

As regards the mechanical features, the most significant are the following.

- The design, service factor, and application track record of the drive-gear. Agitator drive-gears are custom-designed for accepting the shaft thrust and bending moment, for ease of gear ratio change and for avoidance of possible lubricant leakage into the tank below.

- The adequacy of the vertical shaft diameter. This depends not only on the torque, which is easily calculated, but also on the required bending moment, which is not. The bending moment is caused by eddying and unstable flow within the tank; thus it can be promoted by insufficient size or number of baffles. Research articles have been published on the subject, but it is more practical to be guided by relevant experience.
- The impeller mounting and maintainability.
- The materials selection and corrosion and/or abrasion protection of wetted parts.

We stated above that 'usually' an agitator is used for maintaining slurries in suspension, but this is not always the case. Agitators can sometimes be eliminated by careful tank design in relation to the flowrate, and it is always desirable to eliminate equipment items when possible; this is not just to save initial cost – if the item does not exist it will not fail. Typically, pump suction tanks ('sumps') are sized to avoid the use of agitators, making use of tank bottoms acutely sloping towards the outlet. The size and design of these has to be carefully judged against previous applications for the particular slurry and comparable minimum flowrate. It is also often possible, with similar care, to design collecting and transfer tanks to function without agitators. If in doubt about the application, it may be instructive to compute the power input P to the tank in terms of the fluid head

$$P = \frac{V^2}{2} Q\rho + hgQ\rho$$

where V is the fluid velocity in the pipeline, h is the fluid impact height (that is, the height of the pipe centreline above the surface in the tank), Q is the slurry flowrate, and ρ its density. The resulting power input, per unit volume, may be compared with the corresponding figure based on absorbed shaft power per unit volume for similar successful agitated tank applications. The energy of the incoming slurry is not as well utilized as in the case of an agitator, so it is wise to look for about double the latter figure to compensate. A worked example is included in the next chapter.

16.4 Pneumatic conveying

In-plant slurry conveying systems are usually designed by the plant designer (or perhaps a consultant drafted into the team), while the components are purchased as discrete items. Pneumatic conveying systems are more customarily purchased as complete units against a functional specification, or at least the complete system design is entrusted to a vendor who also supplies the principal components. Thus for the purposes of this book, a pneumatic conveying system is regarded as a purchased item, whose acquisition we discussed in Chapter 13. We will not go into much design detail. As for any other purchased item of equipment, the project engineer must verify the available experience of the application, and the specific experience of proposed vendors. In selecting a suitable system and overviewing its application, attention must be paid to:

- all required operating conditions, including the variations of material size distribution and moisture content;
- abrasion of components and pipelines, especially at bends;
- power requirement, which may be unacceptably high to make pneumatic conveying an option; and
- understanding of how the pipelines may become blocked, for example as a result of maloperation, component failure, power failure, and how they will be safely cleared.

As for slurry lines, the line routes should be reviewed at plant layout to minimize the route length and the number of bends.

16.5 Pipework reactions

Experience indicates that a special word of warning is needed on the subject of pipework reactions for two-phase (or three-phase) flow when one of the phases is gas. It is possible to estimate theoretically the reaction on the pipe supports at a bend when a slug of solid or liquid of given size traverses the bend, or alternatively a vapour bubble (a negative mass of displaced liquid) traverses the bends.

Consider for instance a slug of solids of mean density ρ (= 500) kg/m^3 and length l m, travelling at V (= 20) m/s in a pipe of diameter d (= 0.2) m, traversing a 90° bend which has a radius of 1.5d. One can make a rough estimate of the reaction on the pipe by postulating that

Force = change of momentum/time for change

$$\text{Change of momentum} = \text{mass} \times \text{velocity change} = \frac{\pi}{4} d^2 l \rho V$$

Time taken for change = distance traversed/speed = $\frac{\pi}{4}(3d)/V$

Dividing and simplifying

Force = momentum change/time = $\frac{1}{3}dl\rho V^2$

This will clearly not hold for $l >$ length of the bend $\frac{\pi}{4}(3d)$, so if we take

$$l = \frac{\pi}{4}(3 \times 0.2) = 0.47 \text{ m}$$

$$\text{Force} = \frac{1}{3} \times 0.2 \text{ m} \times 0.47 \text{ m} \times 500\frac{\text{kg}}{\text{m}^3} \times 20^2\frac{\text{m}^2}{\text{s}^2} = 6 \text{ kN}$$

The author's belief, based on observation of some failures, is that in fact even greater forces can be generated. Factors which may magnify the reactions include:

- pressure pulsations in the pipeline;
- velocity pulsations; and
- dynamic response of the pipe and support structure to impact loading and periodic oscillation.

It may be advisable (especially for pneumatic conveying) to design for a support reaction equal to the product of the cross-sectional area of the pipe and the maximum pressure. For instance, in the case of the 200 mm line above, if the initial pressure of the motive air is 2 bar gauge (200 kN/m^2), the force is

$$\frac{\pi}{4} \times 0.2^2 \times 200 = 6.3 \text{ kN}$$

One would expect that an experienced supplier of pneumatic conveying equipment would have more experience-based knowledge on which to design the pipe supports, but that expectation has on occasion proved to be over-optimistic. A prudent engineer would do well to take the highest of any of the values calculated above or stated by the supplier, and allow a safety factor of at least 3.0 to allow for dynamic magnification of the load.

Apart from pneumatic conveying, these pipe-bend reaction problems can be of great concern in oil refineries, especially at the outlet of heaters, in particular distillation-unit charge furnaces, where there is always a two-phase flow. The velocities are kept much lower than for pneumatic conveying, but the forces can still be large enough to cause significant oscillations in the furnace outlet bends. The piping here is invariably designed to be rather flexible (and therefore comparatively flimsy) to accommodate expansion. In this case robustly designed dampers may be considered, to allow expansion but prevent oscillation of the pipes. Pipe flanges, which are usually a necessity for equipment isolation, should be located in a position to minimize the bending moment that will be experienced if the pipe oscillates. Usually, the best position is close to the bend.

Chapter 17

Hydraulic Design and Plant Drainage

17.1 Hydraulic design

For transport around a process plant, gravity flow, rather than pumping of fluids, is an attractive option. There is no pump to fail. There are some very simple options for control, for example by regulated overflow over an adjustable weir – an attractive control option in the case of very abrasive slurries.

We have to repeat the refrain that this text does not set out to be a comprehensive guide to specialist subjects, in this case open-channel (or part-filled closed-channel) flow, but the following notes may assist the generalist to avoid some of the worst disasters, in particular, inadequate elevation. Especially in the case of slurry systems, any calculation methods and designs should be backed and supplemented by observation of whatever industry practice is available for the slurry under consideration.

For standard flow calculations in open channels or partially filled pipes, we need to define an equivalent to pipe diameter for the case of a full pipe. There is approximate equivalence of flow conditions if the ratio of wetted perimeter to fluid cross-sectional area is the same, as flow is only opposed by shear at the enclosing wall. The hydraulic radius R is defined as

$$R = \frac{\text{flow area}}{\text{wetted perimeter}}$$

which in the case of a full circular pipe of diameter d yields $R = d/4$. R must not be confused with the actual pipe radius, as it is half the value (in the case of a full pipe).

Turbulent flow in open channels (or part-filled pipes) behaves approximately as flow in full, circular pipe flow of the same hydraulic radius. On this basis we can rearrange the Fanning friction formula (see p. 157) into the format of the Chezy formula

$$V = \sqrt{\frac{gdh}{2fl}} = \sqrt{\frac{gds}{2f}} = \sqrt{\frac{2Rgs}{2f}}$$

where

V = mean flow velocity
f = Fanning friction factor (1/4 of the Darcy factor)
d = diameter of the full, circular pipe equivalent to the channel of wetted perimeter R
s (= h/l) = slope of the channel which drops distance h over distance l

For example, consider the case of a launder 1.0 m wide, running 0.5 m deep, with a water-borne slurry whose velocity V is 2.5 m/s. The equivalent full-flow pipe diameter d_e is given by

$$d_e = 4R = \frac{4 \times (\text{flow area})}{\text{wetted perimeter}} = \frac{4 \times 1 \times 0.5}{1 + 2 \times 0.5} = 1.0 \text{ m}$$

The Reynolds number Re, for the minimum velocity, and based on kinematic viscosity v of 1.0×10^{-6} m²/s is

$$Re = \frac{Vd_e}{v} = \frac{2.5 \times 1.0}{10^{-6}} = 2.5 \times 10^{-6}$$

From friction-factor charts, using a pipewall roughness ratio of 0.0005, the Fanning friction factor

$$f = 0.004 = \frac{1}{2}\left(\frac{h}{l}\right)\frac{gd_e}{V^2}$$

$$\frac{h}{l} = s = \frac{2fV^2}{gd_e} = \frac{2 \times 0.004 \times 2.5^2}{9.81 \times 1.0} = 0.005$$

So s, the required slope, is 0.5 per cent.

This rather cumbersome procedure can be greatly simplified when the liquid is water or a water-borne slurry, provided that the slurry does not contain such a high fines content as to modify its viscosity. In this case the usual practice is to use the Manning formula

$$V = \frac{1}{n} R^{\frac{2}{3}} s^{\frac{1}{2}}$$

or

$$s = \frac{n^2 V^2}{R^{\frac{4}{3}}}$$

The Manning formula can be derived from the Chezy formula by assuming that f is inversely proportional to $R^{\frac{1}{3}}$. The roughness factor n can be taken as 0.012 for reasonably smooth steel pipes when SI units are used. For the case above

$$s = \frac{0.012^2 \times 2.5^2}{0.25^{1.33}} = 0.56\ \%$$

Note that in units of feet and seconds, the Manning formula is expressed as

$$V = \frac{1.486 R^{\frac{2}{3}} s^{\frac{1}{2}}}{n}$$

in which case n is the same as in SI units. However, some textbooks incorporate the constant 1.486 into n, in which case n is different by this ratio. (1.486 is the cube root of the number of feet in a meter.) Clearly, one has to be careful when using tables for n. In any event, the values of n and f (or the roughness ratio from which f is selected) are not very accurately known for most practical cases.

We will now concentrate our attention on water-borne slurry applications; single-phase gravity flow applications are relatively simple to design, and seldom cause problems that are not fairly obvious by application of common sense. The challenge in the case of slurry applications is to avoid blockage by settlement of the solids (calling for high enough velocities), and to minimize abrasion and spillage (calling for low velocities). All of these concerns are assisted by employing relatively straight flowpaths with minimal obstruction.

Often, the design of a plant utilizing slurry-phase transport is based on compromise between, on the one hand, building process stages at successively lower levels and using gravity transport (thereby minimizing the number of pump systems), and, on the other hand, incorporating pumping systems to reduce the overall structure height. As in the case of the design of plants with bulk solids flow, it is essential to ensure that there is adequate height incorporated into the initial design to allow for future changes for whatever reason, including project design iterations. To go with each layout development a hydraulic diagram should be produced, showing the head available and the loss at each flow feature (for example tank outlet, length of launder, inlet to next tank, etc.) in the context of relative elevation.

There are two basic (and interrelated) useful parameters for an open-channel slurry system design: the minimum slope (to maintain slurry suspension) of straight lengths of launder, s_{min} (usually expressed as a percentage), and the velocity head corresponding to the minimum slurry velocity, h_v (expressed in metres). s_{min} is often quoted as part of process technology, and may be arrived at directly by practical experience, whereas h_v is usually derived. The parameters have an approximate theoretical relationship, and the minimum slurry velocity V_{min} is essentially the same as the minimum velocity to avoid settlement in full-flow pipes of comparable diameter, in terms of wetted perimeter.

If the minimum slurry velocity is 2.5 m/s,[1]

$$h_v = \frac{V_{min}^2}{2g} = \frac{2.5^2}{2 \times 9.81} = 0.32 \text{ m}$$

The Manning formula shows that h_v is proportional to s_{min} for the same hydraulic radius. h_v is the *minimum* head required to accelerate the fluid out of a tank or overflow box and into a launder where its required velocity is V. A greater head is required, in practice, to allow for the exit loss. Depending on the detail design of the flowpath, h_v must be multiplied by an appropriate factor. The factor is influenced by features such as sharp corners at the exit from a tank; where practical, it is worth installing half-rounds of pipe inside the tank at a rectangular tank exit, in order to provide a bellmouth effect, or a reducer adjoining the tank at a round exit. The factor required varies from 1.1 for a smooth transition to 1.6 for a sharp-edged exit from tank to launder. (The theoretical values are 1.0 and 1.5.)

---

[1] This would be for a duty such as flow of crystals within a mother-liquor in a chemical plant. Applications such as metallurgical plant slurries tend to require higher velocities because of higher specific gravity difference between the solids and the liquid.

Reverting to the calculation of slope above, a slope of 0.5 per cent is most unlikely to be considered as adequate to prevent deposition of any slurry in practice, because of partial flow conditions. If the flowrate is halved, the hydraulic radius R reduces to

$$R = \frac{1 \times 0.25}{1.0 + 2 \times 0.25} = 0.167 \text{ m}$$

and the value of s calculated from Manning's formula increases to 0.82 per cent. For such considerations, and to make construction tolerances less critical, it is unusual to see a slope of less than 1 per cent in practice; 2 per cent is a more common specified minimum figure. With this comes the consequence that at full flow the velocity V will be much greater, as it varies as $\sqrt{s}$. Not only does the greater V cause increased abrasion, but in the case of an open launder it becomes extremely difficult to prevent spillage at bends, even at very-large-radius bends.

Even for lower velocities, it is very important to take every measure to keep flows straight and to smooth out discontinuities that can cause turbulence. Apart from abrasion, solids are likely to be deposited where there are eddies and can build up to cause blockage or, more usually in an open channel, local overflows.

As an example of the above, and the work of the previous chapter, let us consider a fairly normal plant problem: the design of a splitter box to evenly distribute a single flow into a number of flows going to individual pieces of process equipment. We do this by dividing the flow over a number of identical weirs, see Fig. 17.1. Also included here is an arrangement to keep the slurry in suspension without agitation. For this

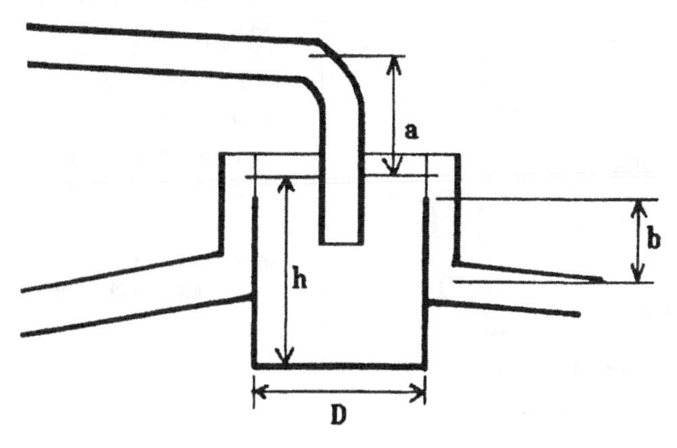

Fig. 17.1 Splitter box

purpose we have noted that in a similar application the slurry was kept in suspension in a box of 4 m^3 volume, with a hydraulic input power of 2.2 kW. In this case the process parameters we will use are: velocity V at pipe outlet = 3 m/s, minimum flowrate Q = 0.25 m^3/s, and slurry density ρ = 1500 kg/m^3, and we have guessed that the pipe outlet may be a distance of 1.0 m above the splitter box water level. Using the formula from p. 191, we see that the hydraulic power input

$$P = \frac{V^2}{2}Q\rho + hgQ\rho = \frac{3^2}{2} \times 0.25 \times 1500 + 1.0 \times 9.81 \times 0.25 \times 1500$$
$$= 4500 \text{ W} = 4.5 \text{ kW}$$

As our Fig. 17.1 splitter box has a diameter D of 2.0 m and the height to the water level h is the same, the contained volume is 6.3 m^3 and the power per cubic metre is 4.5/6.3 = 0.7 kW. This exceeds the value of 0.55 kW for the existing tank, so there should be no settlement, and the margin of excess is not too great, so there should not be excessive splashing. We include also the part of the plant hydraulic diagram for this section in Fig. 17.2. (Note: these diagrams can be and are presented

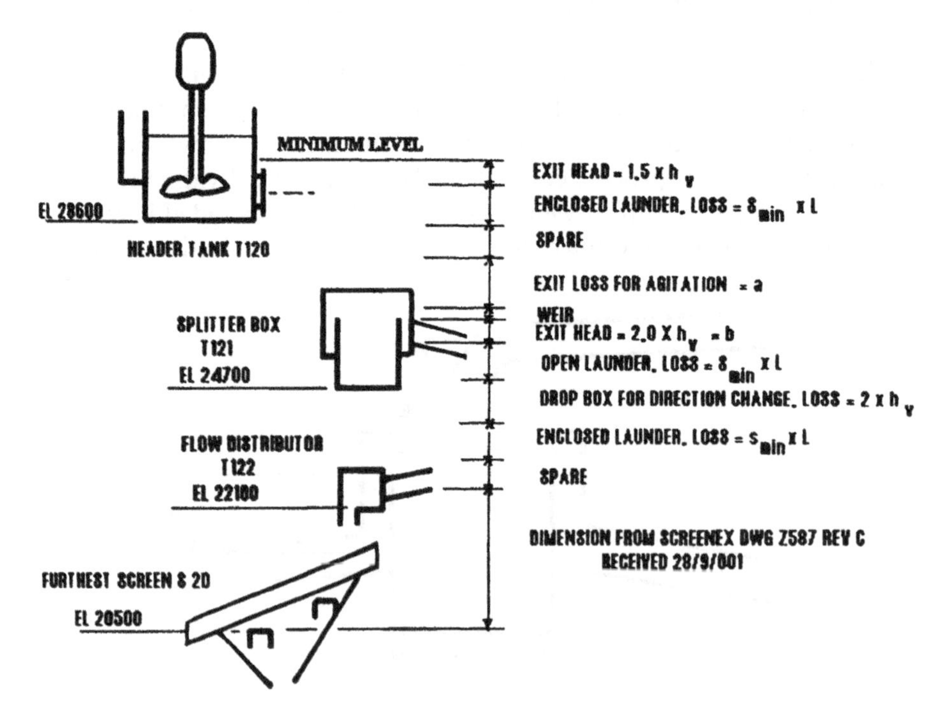

Fig. 17.2 Hydraulic diagram

in a number of ways, depending on the application – the format shown is the author's preference.)

The hydraulic diagram should be fairly self-explanatory in terms of the previous calculations and the resulting plant elevations. Note that at the splitter box outlet, we have used a factor of 2.0 to calculate b. No factor is available from a documented source for this particular overflow box; this comes purely from observations of the existing plant, but one could easily guess such a value by comparison with the theoretical value of 1.5 for a simple abrupt outlet. Note that in these calculations, we have relied where possible on the practice in a similar plant with a similar slurry in order to arrive at a design, which is always the preferred option.

17.2 Plant drainage

Plant drainage is an open-channel flow problem, but not one restricted to slurries. For all too many plant designs, drainage was a matter that received insufficient attention until too late, and consequently it has been one of the hardest problems to rectify. The result is plants with stagnant pools at various places, with the associated safety and corrosion problems and extra labour costs.

Possibly the entire drainage concept of a plant should be included on all P&I diagrams to create the correct mindset, but it is usual to indicate only process drainage (and even that is sometimes omitted). Irrespective of whether it is fully addressed in the P&I diagrams, all sources of process liquid effluent, spillage, and stormwater need to be identified and quantified, and dealt with just as if they were part of the process. A design is required to get each stream to the right place. Where there is a risk of contamination by process substances, there is a requirement for separation of streams, containment of any pollution, and effluent treatment. Typically, a hydrocarbon-processing or chemical plant will have at least three effluent drainage systems: process effluent, contaminated stormwater (stormwater falling inside bunded process areas), and clean stormwater. Drainage of firewater may also be an issue, and should be considered.

The design issues for floor drainage and underground sewers are in fact exactly the same as for in-plant open-channel flow. Settlement of solids may be an issue if these are present in the process. The equations above indicate that, in the case of a floor, because the effective flow area is low in relation to wetted perimeter, a high slope is required and deposition is inevitable at low flowrates. It is usual practice to provide a slope of say 2 per cent, and promote the movement of any deposited materials with hosepipes from time to time.

The same sort of hydraulic diagram as for process streams is recommended, in order to establish *inter alia* the datum of ground elevation for each part of the plant buildings, roads, and surrounding drainage destinations. The drainage diagram should be maintained as a basic plant design co-ordination criterion, and should be reviewed together with the corresponding layout and Hazops.

Chapter 18

Observations on Multi-Discipline Engineering

In the previous chapters, we have mainly discussed work that falls within the responsibility of the process, mechanical, and piping engineers. Before proceeding to a review of the more detailed design procedures which follow on from overall layout development, we will address some of the major responsibilities and concerns of the other engineering disciplines. In particular, we will address some of the issues concerning the overall discipline interfaces, information flow, and planning of work.

18.1 Structural design considerations

A structure is here considered as any assembly of components which supports loads in any direction. We will give consideration to several types of structure, but the most important are the buildings, frames, and foundations which support the plant.

There is a significant advantage brought to a process plant team by a structural engineer who has specific process plant experience over one who has not. The difference includes knowledge of what special problems to anticipate (especially of loading and space occupation), and therefore the ability to ensure that the structural consequences of design decisions are understood by the whole design team, especially process, mechanical, and piping engineers. The anticipated problems must be communicated before it is too late to make changes, that is, at the conceptual stage. An absence of this ability may result in structures which do not properly accommodate the equipment or, more usually, pipes and ducting, and which have braces and strengthening devices that obstruct access and look exceedingly ugly.

Structural design criteria will establish the following.

- Design codes which are statutorily and industrially acceptable.
- Foundation treatment, in terms of a soils report with recommendations.
- Environmental conditions and loads including wind, dust, snow, and seismic design parameters.
- Loading criteria of a general nature, such as superficial loading on flooring and the basis for calculation of frictional reactions at standard expansion joints.
- Standardized components, materials of construction, and systems of corrosion protection appropriate for the project and site circumstances. These must address all the standard components, for example main structural members, fasteners, flooring, cladding and roof systems, stairs, handrails, doors, etc. In addition to structural engineering practice and economic considerations, these choices should be made with the guidance of the process engineers, based on previous plant experience. The choices should take into account corrosion due to process vapours and spillage, and the need for protection against fire and process hazards such as explosions or electrical potentials.

  Corrosion protection considerations should address not only basic material choices and protective coatings but also design details such as avoidance of crevices, and use of electrically insulated connections where required. Special protection should be given where process wastes may build up, say at column bases, where suitable plinths are usually needed.

  The detail design and construction implications of alternative building systems must be considered. For instance, plant buildings made from reinforced concrete frames may appear to be a superficially attractive choice for both economy and corrosion resistance, but the consequences, in terms of longer on-site construction times and relative inflexibility to accept design modification, usually militate in favour of structural steel frames.
- A system to employ standardized design details wherever possible, for example for steelwork connections, expansion joints, cranerail support, column bases, ladders, stairwells, tank platforms, and miscellaneous pipe supports.

Moving on from general criteria to specific structures, the structural engineer must review the basic plant layout, type of structure for each application, and the basic pattern of column spacings and floor elevations adopted (in conjunction with the layout designer). At this stage there must be a critical review of design for horizontal stability. The most economic steel frame design invariably incorporates many braces, but these obstruct the placement of large ducts and pipes and do not accommodate the passage of cranes and plant operators. It is critical to develop basic rules in each building, at the conceptual design stage, as to which bays may accommodate what type of bracing, in accordance with the needs of:

- plant access and maintainability;
- constructability; and
- space needed for ducts, pipelines, and cableracks.

Space reserved for bracing should be shown clearly on conceptual drawings. Portalized structures should be considered when there is doubt about conflicting plant space requirements, or they may be dictated by crane travel path.

When it is decided to enclose any space, the basic concept for the ventilation system must be determined in so far as it affects the structural design (usually, the type of roof). Note that the ventilation requirements are very much process-related.

18.2 Plant and equipment loads

It is possible to take the view that the loads to be supported, when not dictated by the basic structural design codes and building regulations, will be advised to the structural engineer by the suppliers or designers of the supported equipment. It should certainly be the structural engineer's intention to obtain the vendors' certified loading information in every case. However, it is also necessary to check very critically this information, and ask the right questions. Some of the problems typically encountered include the following.

- Rotating or reciprocating equipment, such as fans, vibrating screens, and centrifuges. These of course have to be assigned a dynamic as well as a static load, and the structure must be designed to avoid any natural frequencies near the operating frequencies. Possible loading at start-up, shutdown, and fault conditions should be ascertained.

  Inherently unbalanced equipment mounted on flexible supports, such as vibrating screens mounted on springs, often operate at above their critical frequency. The dynamic load transmitted to the supporting steel when passing through this frequency will be much greater than under normal operating conditions. Usually – but check with the vendor – this has little effect when accelerating quickly up to speed, but there may be a much greater problem when stopping by disconnecting the power supply. Insist on receiving full information, and check whether brakes are necessary.

High-speed equipment, such as fans, may develop high, unbalanced loads in the 'dirty' condition (when process material builds up on the impeller). There should be a clear understanding of how this will be limited, for example by the provision of a vibration sensor coupled to an alarm or cutout device. Alternatively, it may be better to put the equipment at ground level – a decision which should obviously be taken at the conceptual design stage.

Special care is needed to deal with equipment in which the driver and driven equipment are separately mounted. Usually the relative alignment must be critically maintained; a 'flexible' coupling sometimes accommodates only fractions of a millimetre without unacceptable wear. A standard electric motor will develop more than double its full-load torque at start-up, and this is usually a suddenly applied load, requiring an impact factor.

- Belt conveyors. The belt tensions must somehow be accommodated by the structure, and can either be taken through the conveyor gantry or be taken out at the headframe. The design to accommodate belt tension must be integrated with the design of the expansion joints to accommodate thermal expansion. Design belt tensions should be based not just on the normal operating condition but also on drive motor size and maximum torque at start-up, or under a possibly worse condition a following a motor trip under maximum load, and operation of the holdback. It should be possible to obtain figures for each scenario from the manufacturers of the motor and holdback respectively.
- All space-containing equipment, vessels, pipes ducts, or even floors which normally contain fresh air, gas, or light liquid, may at some time or other be full of water or solids. The situations particularly to watch out for include hydraulic test, precipitation of dust and process solids, blockage of chutes, overflow of vessels and bins, and spillage from belt conveyors. Such situations have to be discussed with the mechanical and process engineers to arrive at appropriate design criteria. It may be uneconomical to design for the worst case. For instance, in many plants it is considered acceptable to design

dust-extraction ducts for a maximum of 30 per cent volume of settled dust; such criteria may become plant operational parameters, and should certainly be communicated to plant operators.

- Normal piping and ducting expansion joints transmit no axial loads. Thus if there is pressure (or vacuum) in the pipe the ends will be forced apart (together) and these forces have to be accommodated by the supporting structure.
- When overhead travelling cranes are installed in a building, special standards of alignment and rigidity are required. These should be specified by the mechanical engineer in conjunction with the crane specification, or specified by the crane supplier.

18.3 Civil engineering

The civil works associated with a major process plant often represent a challenge which differentiates this work from other civil engineering applications, especially in the case of solids-handling plants, which tend to be individual in layout, non-modular in design, and include many special foundations. Information flow is the essence of the management of engineering, and much of the discipline design information is available last (being dependent on layout and design development for the supported process equipment and structures). And yet, by the sdictates of gravity the civil construction must be completed first.

This 'pinch' position within the project schedule inevitably means that there is great pressure to bypass the logic of the critical path by whatever short-cut means are available (as discussed in Chapter 27), which means working under enhanced levels of uncertainty, change, and therefore stress. It also dictates that a high level of management skill and cool judgement under stress are needed for the leadership of this discipline, the lack of which can ruin a project as effectively as any shortcoming in the core process skills. Needless to say, meticulous planning of work is essential.

There is clearly an overlap between civil and structural engineering, certainly so in terms of the definition of a structure quoted above (Section 18.1, p. 203), which embraces all load-bearing constructions. How these two disciplines are integrated or co-ordinated depends on the skills and inclinations of the project team. For the purposes of this book (and in many project teams) there is no real separation, and parts of what is quoted under one heading are applicable under the other. As a generalization, the 'structural' engineer deals with steel structures which are usually shop-fabricated and require only assembly on site, whereas 'civil' works are practically all performed on site and include reinforced concrete structures.

Front-end civil activities include those associated with site selection: survey, gathering of sub-surface information and hydrological data, and analysis of their effects on the comparative suitability of available site options. This is likely to be combined with infrastructural and environmental studies beyond our scope. Arising out of these considerations will be a decision on site selection corresponding to a basic plant layout, with an understanding of the basic site topography, soil conditions and foundation treatment required, and the way to develop the site in terms of terrace levels, access roads, and drainage. Impacting as it does on the construction estimate, this work will normally be concluded before the decision for project implementation, but the conclusions may have to be refined as part of the initial project implementation work. Usually there is very little time for this, as the civil engineer is quickly under pressure to release the information that will facilitate the planning and contracting of site clearance, bulk earthworks construction, and the setting out of construction facilities.

Re-emphasizing the commentary made in Chapter 17 on the subject of plant drainage, it is essential (before releasing bulk earthworks and plant infrastructural designs for construction) to ensure that the layout and elevations employed correspond to a comprehensive and acceptable plant drainage design.

Civil works are a significant part of both the cost and construction time of a plant, and consequences to the civil work must be adequately considered during preceding engineering work by other disciplines, in particular, during layout and when purchasing large equipment items. The designs of foundations and bases for large machines usually incorporate features and dimensions which are dictated by the machine vendor. Sometimes the complete detailed design is obtained from the vendor, but usually the construction of the foundations is not a vendor responsibility, as it is not considered economic to fragment the site civil construction work.

Whether the machine foundations are designed by the vendor or by the project civil engineer in accordance with the vendor's requirements, there may be features, details, and tolerances requested which are difficult to achieve, expensive, and unnecessary. These circumstances are apt to come about as a consequence of the way the machines are procured; the machine supplier may have little incentive to concern himself with the problems of foundation construction – on the contrary, he may find it quite convenient to demand what can hardly be achieved. Such possibilities are easily minimized, once they are recognized, by including adequate requirements within the purchase specifications for the machines. For instance, require that the machines are constructed to accommodate specified foundation tolerances, because it is usually possible to provide for greater adjustment within machine baseplate design when there is a demand. If typical foundation details are available for review by the civil engineer before ordering such machines, so much the better.

Following on from the usual decision to go ahead with civil works before the completion of all other discipline designs that affect the work, there are invariably a number of 'miscellaneous support plinths' which become necessary after civil work in the corresponding plant areas is complete. These may be for pipe supports, miscellaneous electrical panels, additional access structures, small machine auxiliaries, and so on. There may also be changes to floor slabs and drainage details to allow for plant access once the small details of plant configuration have been finalized. At the conceptual design stage every effort should be made to minimize the disruption and unplanned work caused by such additions by the other disciplines involved; however, the need can seldom be eliminated, and will be greater in the case of shortened project schedules. An idea of the quantity of such additions should be possible from previous experience, and even if the quantity is a complete guess, it pays handsome dividends to plan for this eventuality. Draw up 'standard designs and methods' for such retrofit work, and include (on a provisional basis) a quantity of such work in bills of quantities and work budgets and schedules. The impact can be quite considerable, but if the work is officially planned it is better accepted.

18.4 Electrical engineering

Process plants are for the most part dependent on electric power for motive purposes, to drive the process equipment and the materials transport systems. Other types of driver, in particular the steam turbine, have sometimes been employed, especially for major drives such as turbo-compressors (with which a turbine driver has a natural compatibility in terms of speed and speed variation), and in plants where steam is generated for process consumption or as a by-product in waste-heat boilers. Gas engine drivers (for reciprocating compressors) and gas turbines (for turbo-compressors) are also often employed for oil-or gas-field work. Since the introduction of programmable logic controllers and computerized plant control, and the advances made in variable-speed AC drive technology, the use of electric motors for all drives has become increasingly attractive. Any available steam is more likely to be used to generate power than to drive machines directly, except perhaps for large turbo-compressors, and the use of gas turbine drivers is likely to be confined to the more remote fields.

Electric power reticulation is one of the vital systems of the plant, but (with important exceptions which will be addressed later) does not interface directly with the process as does for instance the mechanical equipment, the instrumentation, or the piping. The power reticulation serves the equipment, which serves the process. Electrical engineering work is usually the responsibility of a separate person, a project electrical engineer, and this relative distancing from the process has to be borne in mind as a possible source of miscommunication and sub-optimal design or error. Like every other part of the plant, there are aspects of the electrical design which must closely match process requirements and must also allow flexibility for changes, which may appear as part of the iterative design process and as post-commissioning development. We will address in particular these interface aspects of electrical engineering; subjects such as the design of high-voltage systems are not within our scope.

The start-point for electrical design is the electric motor and other consumers list. This list is typically drawn up by process and mechanical engineers involved in the selection of the driven equipment. The electrical engineer needs to probe this data and understand how it was drawn up, with special reference to the expected load factors (namely, ratio of actual kW or kVA to the value used as design basis). Rarely, drive sizes are underestimated. More usually, they are consistently overestimated as a result of the practice of putting 'factors on factors'. There may be a factor of conservatism for increased process performance requirements, a factor for driven-machine performance tolerances and deterioration of efficiency, and then perhaps a factor to allow for uncertainty, and, then again, rounding *up* to the nearest standard motor size.

Whereas individual motors tend to be oversized rather than under-sized, it is more common to make insufficient allowance for increased numbers of drives, both as project designs develop and for the future. When deciding on criteria for future additional drives, it is often more illuminating to review past experience of project design and plant development than to rely exclusively on the process technologist.

Sometimes the detailed electrical design commencement is attempted too early, when it is simply a waste of effort in terms of ever-changing drive requirements, and maybe only the essential longer delivery items, say the main transformers, should be committed for early purchase. This of course implies that the transformers may be oversized; few engineers would risk undersizing them. However, the margin of uncertainty and therefore additional rating should be less than that which may be applied to individual drives, because an averaging factor can be applied to the potential growth of the various individual drives. Also, spare transformer capacity is anyway needed for future additional drives; it is not as detrimental to cost, power factor, and efficiency as the consistent oversizing of individual motors.

Included on the motor list are likely to be some frequency-controlled variable-speed drives. Starting and operational torque–speed characteristics of these drives must be understood, and the adequacy of both the drive motors (whose cooling at low speeds may be a problem) and the control gear characteristics should be verified for the service demands. There is more normally a problem for drives with constant torque characteristics, such as feeders and positive-displacement pumps and compressors, unless the latter are equipped with unloaders for reduced-torque starting. High-inertia loads – big fans and centrifuges are normally the worst – must also be identified, and starting methods must be designed in accordance with the electrical system characteristics. In addition, it may be decided to employ electrical 'soft-start' facilities on certain drives that are subject to prolonged start or operational stall conditions, rather than employ a mechanical device such as a fluid coupling.

The next fundamental design document for plant electrical engineering is the single line diagram, very much the electrical equivalent of the flowsheet. This document is not the exclusive preserve of the electrical engineer; it requires the input of the process and mechanical engineers. Its format is dependent on the following.

- The layout – especially as regards the configuration of satellite substations.
- Prioritization of power, securing supplies to important consumers under fault conditions, and emergency power supplies to critical users, for example by provision of standby generating capacity.
- Possible future changes, in particular future plant development requirements, and uncertainties current at the design stage.

The routing and support of electrical cables and cable-trays, and the location of miscellaneous electrical items within the plant, are important interface issues. Plant design criteria must include the establishment of basic practices of how cables and cableracks will be run, for example, underground or mounted on structural steelwork, with racks in the horizontal or vertical plane. Basic sizing and routing of cableracks should be included in the plant layout before freezing it, and further development of cable and cablerack design should be closely integrated with piping design, often with a common structural support system. Instrumentation cable and cablerack location should be co-ordinated with the electrical work.

The interface between the plant electrical system and the control system varies between plants and has to be defined. Sometimes the interface corresponds to the owner's plant maintenance practices, in respect of which maintenance trade has responsibility for and access to which equipment, for instance programmable logic controllers (PLCs). An important aspect of this interface is the facility for emergency stopping of machinery. It is possible to mis-design the relationship between the control and electrical systems such that a computer failure can disable the local emergency stopping facilities – not a happy prospect. There have also been instances where the 'scan' time of the PLC exceeded the contact period of an emergency stop button. If there is any doubt whatever, emergency stop devices should be hard-wired.

Finally, there are sometimes aspects of plant electrical engineering in which the electrical design will interface directly to the process, most commonly in electrolytic processes. Here there is no mechanical equipment intervening between the process and the electrical supply. The flow of information within the project team is different. This must not be allowed to be the cause of error or omission, such as failure to completely specify the electrical characteristics of the cells, the back e.m.f. and its significance under shutdown conditions, and the insulation requirements for busbars and potentially energized and accessible equipment.

18.5 Instrumentation and control

Instrumentation, and more particularly the associated methods of information presentation and processing and plant control, have probably developed more rapidly than any other facet of plant engineering in the latter part of the twentieth century. At the time of writing there seems to be just as much scope for rapid future development, with particular regard to increased electronic intelligence. And yet many of the points of issue within a process plant design remain the same; we will try to summarize the most persistent of these, most of which relate to inter-disciplinary information flow and co-ordination.

The plant instrumentation and control requirements are directly linked to the process design as portrayed on the P&I diagrams. There is therefore a natural direct route of communication between the process engineers and the instrument engineers regarding the overall measurement and control system functional design (the 'control philosophy') and the individual instrument data sheets describing the process conditions and measurement range and accuracy required. However, the instrumentation and control systems also serve mechanical equipment in non-process-related functions, such as machine protection and control of machine auxiliary services. This latter requirement, dependent for the most part on information from mechanical engineers and machine suppliers, must also receive adequate attention when scoping instrumentation and control work and devising co-ordination procedures.

The instrument/mechanical interface is not just a matter of ensuring that machine instrumentation is accommodated within the overall control system. It is necessary to work in the opposite direction, to ensure that all instrumentation and control features of machinery are fully compatible with the overall plant systems, including their functionality, interface parameters (such as signal and electric power characteristics), and component standardization. This requires a direct participation of the instrument engineer in the specification and selection of most mechanical equipment packages. The instrument engineer must develop a complete understanding of the functioning of the mechanical plant and its critical needs, for example, signals which are essential to plant safety and must be configured in a 'fail-safe' manner.

We previously mentioned the importance of co-ordination of cable and cablerack design with electrical piping and structural design, with the main objectives of preservation of plant access, avoidance of clashes, and the integrated design of supports. 'Clashes' in the instrumentation context should include any possible electromagnetic interference due to the proximity of high-current cables and magnetic fields. It is also important to identify the space, support, and access needs for local instrument panels, marshalling boxes, and the like, and include them in integrated layout development work just like an item of mechanical equipment. Failure to give these items adequate attention frequently results in a 'retrofit' design, whereby instrument panels obstruct access or are located where damage from moisture, vibration, or maintenance activities can be expected. Design co-ordination in this regard can be facilitated by drawing up appropriate layout standards to be utilized at the overall plant layout development stage.

The installation of each instrument constitutes a design interface and, again, simply retrofitting the instrument installation to a finalized vessel or piping design is not good practice. Standards or special designs need to be developed and be available before detailed vessel or piping design is commenced, including recognition of the hardware such as connecting flanges, dip-pipes, and instrument-isolating valves, and whether these items are the responsibility of the instrumentation or interfacing discipline. Mostly, this information can be shown as typical details at P&ID level.

There are also important discipline interfaces in the case of control valves, motorized control devices (such as electrically actuated valves), and safety and relief valves. The interface is easily managed if appropriate procedures are available from the start as to which discipline will be responsible for what.

Some instruments have layout implications that have to be recognized at an early stage: flow-measurement devices usually require straight lengths of associated piping; analysers of various kinds may require quite elaborate interfacing to get a representative sample of process fluid, and to dispose of the sample and any associated waste. The spatial consequences of the design interface, and recognition of all the components, need to be clarified at the conceptual design stage. The best way to notify this information to all disciplines is on the P&I diagrams and layouts.

It is possible that the multi-discipline implications and costs of installing a particular type of instrument – if properly appreciated at the time of developing the control philosophy – should result in other solutions being sought. There was for instance the case of a radiometric analyser which appeared as an innocent little symbol on a plant P&ID. When the item was purchased, along came the information that it must be surrounded by a 1 m radius of representative sample of (solid) process material. In consequence, the innocent little P&ID symbol ended up as including a large bin with hydraulically operated extraction devices to maintain an acceptable level, all occupying two additional floors of process plant building. The instrument engineer was not popular.

Instrument access requirements for maintenance and calibration have implications which can also be far-reaching and must be appreciated at

Fig. 18.1 The instrument that grew to occupy three floors of a process building

the plant layout stage. Generally it is not acceptable to rely on access by mobile ladder or scaffold, and it is a question of whether access by cat ladder or stairs is needed and how the fixed platform may relate spatially to the instrument. Instruments whose measurements impact on product sales, such as petroleum product flowmeters, can require plant additions such as calibration pipe loops, which take up significant space.

Many instrumentation systems and control devices require power sources, electrical or pneumatic, which have to be reticulated around the plant. The cost of 'instrument air' is routinely underestimated, mainly because the associated piping costs are not appreciated when the P&I diagrams are drawn up, but also because of a disinclination on the part of the instrument system designers to make any compromise on air-quality requirements. Even a few parts-per-million of oil may be regarded as unacceptable, with consequent cost increase to the air compressor. Both instrument power supplies and instrument air have to be provided in an adequately secure fashion in relation to the consequences of their failure.

Instrumentation is generally the most sensitive part of a process plant, and may require special enclosures (which can be demanding on plant-space and restrictive to access) and air-conditioning, both factors which need to be evaluated at the plant conceptual design stage.

In consideration of some of the issues raised above, it should be noted that the decision to incorporate each instrument in the plant design can have far-reaching implications for the performance of other disciplines. It is easy to grossly underestimate the cost of an instrument, because so many of the cost implications – access platforms, air-piping and compressors, vibration-free mounting, etc. – show up in other disciplines. This is the case even without going to the extreme of the miserable radiometric analyser that doubled the size and cost of a plant building! The accuracy required of a process measurement can have equally far-reaching effects. For example, a high accuracy requirement may dictate that the weight of a bin's contents must be measured by a load-cell system rather than by strain-gauges on the support structure, requiring a much more costly structural design concept. Or high accuracy may dictate that a weigh-flask system must be used rather than a belt weightometer, with major layout and cost implications.

In conclusion then, it is important at the conceptual design phase to critically review the proposed plant instrumentation, ensure that the multi-disciplinary consequences have been considered, and (in this light), concerning expensive items, ask the following questions.

- Is the instrument worth its total cost?
- Are there cheaper acceptable ways of achieving the required measurement or control?
- Does the measurement need to be as accurate as specified? (When this has important cost consequences.)
- Can the instrument be more economically relocated?

Finally, there are a few instrumentation-related items which tend to be omitted from scoping studies by default of 'discipline ownership' and are therefore worthy of mention. These are:

- plant fire detection systems (sometimes omitted because they are not directly process-related and are therefore omitted from P&IDs);
- plant information management needs which are not directly process-related, such as those for maintenance;
- communication systems; and
- sampling systems, which can have major layout and cost implications, and can easily become whole buildings.

Chapter 19

Detail Design and Drafting

In the discussion on layout development, we necessarily started with some of the broader aspects of plant design and drafting, and in the commentaries of various aspects of engineering, we concentrated on some broad issues. In the following, as we proceed to more of the detail, it will be seen that the details and, surrounding them, the systems and standards which govern the development of details, are in fact an inseparable part of the higher-level designs and design procedures. The way in which the more critical details are to be designed requires attention at the layout stage (usually prompted by the overall design criteria), and, for a well-engineered project, the details link inseparably to the way in which equipment is purchased and construction is managed. In other words, consideration of the design details cannot be an afterthought; they have to be anticipated in the sense that they must fit into a system of fundamental order.

There is no such thing as an unimportant detail in process plant design. The chain-like nature of a continuous process means that the failure of one individual item often leads to the failure of the whole process. This may at first seem to be a contradiction of earlier statements about the importance of recognizing and concentrating attention on critical items, and applying limiting return theory to the rest. This is not the case if it is understood that the non-critical design items are only non-critical because it is relatively easy to get them right, and they will be right if developed within a suitable system. The system of design is therefore always itself a critical item.

We will begin by considering some of the more narrow discipline issues, in order to build a base from which an overview is possible. The details are addressed mainly to promote an understanding of the methodology of design system development.

19.1 Structures

As stated in the previous chapter, it is preferable to design all elevated equipment supports, plant enclosures, access platforms, piperacks, and pipe and cablerack supports as steel structures. Exceptions typically include:

- supports for major machinery items, where heavy dynamic loads make it advantageous to mount the items on elevated, heavy concrete structures or blocks integral with the foundations;
- areas where the need for fireproofing makes steel structures less attractive;
- areas where elevated concrete flooring is needed for collection and disposal of liquid and solid spillage, or for protection from fires beneath;
- areas where reinforced concrete is preferred for corrosion resistance (sometimes unjustifiably!); and
- areas, such as electrolysis houses, where some degree of electrical isolation is required.

For the rest, even where reinforced concrete construction may at first appear to be a more economical option, it is usually avoided because of the relative difficulty of modifications and the increased quantity of site work: steelwork fabrication can proceed in the shop while foundations are constructed.

Steelwork is designed in the format of line diagrams, based on the layout drawings and equipment loadings. From these, steelwork arrangement and detail drawings are prepared, showing also the detailed connections to adjoining parts (column bases, supported equipment, etc.); the arrangement of flooring, handrails, kickflats, stairs, building cladding, and the associated purlins, and architectural details. At this point the drawings are typically passed to a steel detailer who is employed by the steelwork fabricator. The detailer designs individual connections, for example welded gussets and bolted joints, and produces detailed manufacturing drawings for each steel member, bulk material cutting lists and diagrams, and a corresponding numbered assembly drawing to facilitate manufacturing control and erection. The responsible structural engineer checks that the details are structurally acceptable; invariably they have to conform to pre-ordained typical details.

Sometimes the workshop detailing is carried out by the project steelwork design team or a separate sub-contractor, either because such a work breakdown is preferred or because the steelwork fabricator does not have, or is adjudged not to have, adequate detailing competence. This arrangement has the advantage that steel detailing can commence before the award of the structural steelwork contract, but has the disadvantages that:

- the detailing may vary from the customary workshop practice employed;
- there may be some duplication of work, especially if the steel details have to be linked to numerically controlled machinery inputs; and of course
- the IFC component of project cost increases.

There are software packages available that embrace the structural design, arrangement, and shop detailing of steelwork. As for all software, the potential user has to assess carefully whether its employment is really justified, considering not only the cost of software and its obsolescence but also availability of trained users and possible consequences (such as difficulty in making changes). For example, the program may deal with the entire structure as a monolith, such that changing one member may result in unnecessary changes to all the other member drawings.

Detailed steelwork design, and its effect on access and relationship to connected items, can be some of the most challenging aspects of plant design, in particular for solids handling plants, which sometimes need complex and non-perpendicular connections between conveyors, chutes, and supports. Information flow is often problematic: steelwork connects to, and supports, most other discipline items. Especially on a fast-track project there may be a need for many site modifications, and still there may remain some unsatisfactory compromises.

The draughtsman's task may be greatly eased, and the amount of errors reduced, by using the maximum amount of standardized designs and systems of design. This practice also enhances constructability and plant appearance. Standardized designs should include:

- steel connection details;
- all flooring systems employed, including sizes of panels, intermediate support systems and details, cutouts around equipment and vertical pipes and cableracks, proprietary details (say for grating and its securing clamps), arrangement and details of handrails, stanchions, and kickflats, and connections to main structural members;
- similar details for roofing, side-sheeting, doors, windows and translucent panels, and ventilators;

- stairways, ladders, and platforms;
- crane girders and their connections to columns and to the cranerails, and other lifting beams and devices (davits, trolleys, etc.);
- column bases which should preferably be symmetrical wherever possible (otherwise a certain percentage invariably has to be cut on site and rotated 90 degrees to suit the foundation bolts);
- bracing systems, including a few choices to suit enhanced passage of pipework and personnel; and
- steelwork expansion joints and sliding supports, for various applications and loading situations.

In conjunction with the interfacing disciplines, standard arrangements and details are needed for:

- equipment and vessel supports and holding-down bolts;
- piperacks, and the configuration of piperack junctions, road crossings, and piperack/building interfaces;
- miscellaneous pipe, duct, and cablerack supports, plus supports for instrument panels, motor starters, and so on; and
- access platforms, per type of equipment, vessel, instrument, or grouping of valves. (Standard support steelwork and access platform arrangements should be worked out for commonly encountered applications, such as tank agitators, for which the equipment access and maintenance is important.)

General plant design practice is to lay everything out within a perpendicular grid system, avoiding oblique interfaces. This practice is sometimes sub-optimal in the case of solids handling, but then special care is needed when designing the non-perpendicular interfaces to reduce the possibility of error. A general principle to facilitate design in such situations is that there should be no more than one non-perpendicularity per interface. Interfaces which have attributes that are non-perpendicular in both plan and elevation should be avoided.

This is best illustrated by the case of an elevating belt conveyor interfacing with a building to which it is not perpendicular in plan. It is generally a false economy to try to interface the conveyor gantry and headframe directly to the building. Better practice is to mount on the building an intermediate frame parallel or perpendicular to the conveyor, and design the conveyor headframe, gantry, drive platform, etc. as a separate structure. This may be mounted on the intermediate frame by sliding supports if needed, but in any case can be adjusted to fit along its length without the complication of corresponding lateral adjustment. The end result may look as if the intermediate frame could have been designed as part of the conveyor headframe, but the mental liberation provided by regarding the intermediate frame as a separate entity seems to have a positive benefit for both design and construction.

19.2 Piping

In Chapter 14 we dealt with the fundamental aspects of piping engineering, by which are set up a system of pipework specifications which identify all the components (pipe, fittings, valves) within each pipeline. We also discussed the need for basic design criteria which included the means of dealing with thermal expansion, the codes for determining the acceptability of stresses, and the need for development of standard piping configurations appropriate to the process. We will now develop these issues further into their practical application.

Firstly, here is a checklist of the basic requirements that should be in place before starting detailed piping design.

- P&I diagrams (certainly finalized to the point of including all but a few small-bore lines) and the corresponding line list information.
- Piping design criteria, including a library of pipeline material specifications, and the minimum contents outlined in Chapter 14.
- Plant layout drawings. There is an overlap in the following text between layout drawings and the detail work, because the layouts have to anticipate some of the details now discussed.
- Drafting system (part of overall design criteria).
- Piping materials management system, including a fully developed interface governing the catalogue of piping components, and the system for take-off of parts and their roll-up and communication for purchase and construction.

Nine basic observations on the design of individual pipelines follow.

- The pipes must fulfil the requirements of the process, in particular, the slope, drainage, and venting arrangements must be acceptable. The piping must be correctly configured for in-line instrumentation and sampling devices.
- Most piperuns are horizontal, or at a small inclination to the horizontal as required for drainage. By their nature, therefore, pipelines tend to block off plant access, and putting the pipes in a

position where they do not obstruct access is the main challenge to the piping designer.

- Suitable facilities must be provided, if required, for routine maintenance (for example flanges for dismantling).
- Suction lines must be configured correctly (see Chapter 14).
- In the case of slurry lines, the requirements outlined in Chapter 15 must be observed.
- The pipes must be adequately supported by a system of supports which itself poses minimal access obstruction.
- The pipelines must be configured such that neither stresses nor movements caused by thermal expansion exceed the allowable values, and that consequent loads on supports or connected equipment do not exceed allowable values. If expansion joints are used, any unbalanced pressure loads must be acceptably absorbed.
- Access must be provided to valves and in-line instruments, preferably at ground level or from existing operating floors, or otherwise from special platforms which do not themselves become a source of access restriction.
- The pipelines must be kept as short and straight as possible, while fulfilling the other needs.

Arguably the most important consideration when laying out the route of a pipeline is its support (except for slurry duties, where directness and elimination of bends often takes priority). Supports must be spaced at adequate intervals to prevent the pipe being over-stressed in bending or deflecting excessively and allowing stagnant pockets to form. Systems of design should include standard tables for minimum support spacing as a function of pipe dimensions and fluid. Special care is required in the case of plastic pipes, as previously stated; there is often a vendor standard for these. It is essential to think out the support system when designing the line, and not leave it until later. Failure to do so usually results in some inadequate supporting, or supports which are an unacceptable obstruction or are very complicated and expensive.

Considering the difficulties of co-ordination faced by a multidisciplinary design team, and the need to minimize design iterations by foreseeing the whole eventual plant when preparing initial layouts, pipe support systems should be substantially set out for batches of lines from the outset; for example, pipes should be run on racks whenever possible.

Moving on to the adoption of procedures and standard designs by which the objectives set out above are facilitated, we start with the perspective of the whole plant, which may be divided into process areas (or units), in-between-process areas, and offsite areas. The most common solution to running pipework between process areas is to mount it on elevated piperacks, permitting access underneath. Offsite, it is normally acceptable to run the pipes on sleepers or low supports at ground level, and make appropriate arrangements (road bridges or conduits) for crossing roads. Such arrangements may also be acceptable in certain cases between process areas. The configuration in plan of the piperacks and sleeper racks (to minimize their length), is a fundamental layout issue which we will not revisit here; we will move on to addressing the vertical planes.

Basic layout considerations will have established which pipes run on which piperacks, and to these should be added an allowance for future development and an understanding of what other services may run along the piperacks, for example cableracks. The next step is to establish the sectional arrangement of the piperacks and their contents, noting the following piping groups which must be especially catered for:

1. pipes which must slope;
2. hot pipes, which require insulation and provision for thermal expansion;
3. pipes which require access for cleaning or occasional dismantling, including ducts which require entry for cleaning or refractory maintenance, and high-maintenance lines (such as slurry lines) requiring occasional replacement;
4. plastic and small-bore pipes, which require special supporting arrangements.

The first group is the most difficult. It is seldom a viable proposition to slope the whole piperack. If there are only a few pipes that slope, it is sometimes possible to run them on the outside of the piperack. The more usual solution is to suspend them by hangers of appropriately varying length. If the runs are too long it may be necessary to make arrangements for one or two intermediate drainage points, and corresponding reverse slopes or risers, to avoid the pipes from becoming too low (or starting too high). In any event, these issues must be addressed at the outset of piperack layout.

We will assume in the following that thermal expansion must be accommodated by designing sufficiently flexible pipe configurations – expansion joints are usually not permissible because of reliability concerns, deposition of process materials in the pockets of the joints, and the difficulty of catering for the end-thrusts created.

All drawing offices involved in the design of hot piping need to have tables of standard configurations of pipe loops and bends, and perhaps a few more complex shapes. The tables should contain formulae or graphs to determine the values and acceptability of the movements, stresses, and anchor loads, depending on the temperatures, materials, and pipe-sizes employed. These are all that are needed to design most hot piperuns, dividing the pipelines up into sensible sections which give attention to anchoring, guides, sliding supports or shoes, and adequate end-clearances, all of which should be covered by drawing office standards. Code requirements or common sense may dictate that the designs should be subsequently analysed by formal flexibility calculation and stress analysis (invariably now by the use of proprietary computer programs), but except for the more intractable applications, this exercise should be a confirmation of what has been adequately designed in the drawing office.

The above process will yield a sensible system of configuring the pipelines into bends and loops that are sufficiently flexible. The ensuing preliminary pipeline designs must then be married to the piperack layout.

Junctions between piperacks, and external connections to piperacks, are facilitated by a change of elevation. Where there are a number of pipes lying alongside on the rack, clearly only the outermost pipe can receive a horizontal connection without obstructing the other piperuns. It may pay to put a large commonly intersected line (such as a cooling water header or common vent or flare line) on the outside for this purpose, but in general, it is convenient to intersect lines vertically, and therefore to intersect piperacks at different elevations, to prevent obstruction (see Fig. 19.1).

In the case of pipe loops it is sometimes possible to arrange all the hot pipes on one side of the rack, and to accept that all the hot pipes will be looped at the same point along the piperack, in which case the loop may be in the same plane as the rack pipes. Otherwise, and if there is any doubt about possibly conflicting design development or future needs, the loops should be stationed in a horizontal plane above the rack pipes, and the pipes connected vertically to the loops.

Moving on to the next two pipe groupings, the arrangement of access when required for cleaning or dismantling should need no elaboration. Plastic and small-bore pipes require an intermediate support system. One design solution is the use of cable trays to support the pipes along their length.

Fig. 19.1 Piperack junction

Finally, it is necessary to provide ergonomically positioned stations for valves (including relief valves) and instruments, together with the associated access platforms. Isolation valves in piperacks are usually best situated at individual process unit battery limits, such that during unit shutdowns all lines to a unit may be isolated all in one place. Attention should also be given to the positioning of break flanges (where disconnection or the installation of spades may be required) and to the permanent installation of spectacle blinds, but the design criteria may allow these to be accessed when needed by temporary scaffolds.

Following the steps outlined, the best overall three-dimensional piperack layout can be chosen, followed by detail design of the steelwork and the pipework.

Piping design within process areas is inevitably equipment-centred and, in the case of a predominantly fluid state process, largely determines how the equipment items are orientated to each other. A basic layout study needs to be made for each equipment item, and standard layouts should be available for most common equipment items, including each type of pump and compressor, and banks of heat exchangers. To some degree it may be possible to purchase equipment – especially pumps and compressors – such that the piping connections are suited to the required piping layout, and this should be considered before finalizing the appropriate equipment data sheets.

In any event, the piping arrangements around individual equipment items should be verified against the constructional, operational, and maintenance needs, many of which may not, or not yet, appear in the P&I diagrams. This is especially the case for those features which arise out of the equipment vendor's information. Examples of such special needs include:

- the provision of isolation flanges (usually with sufficient flexibility for the installation of spades), spectacle blinds, or removable pipe spools in order to facilitate equipment inspection, cleaning, maintenance, hydro-test, and removal;
- the provision of special supports to support the piping during disconnection for maintenance or for the insertion of spades;
- installation of temporary strainers for flushing during plant precommissioning; and
- the provision of equipment venting and draining facilities.

Most of these requirements should appear on the equipment vendor's drawings, but there should also be a design review by the mechanical and process engineers (and possibly by the vendor), especially for the more complex pieces of machinery.

The detail design of vessels, including tanks, columns, reactors, and general pressure vessels, is invariably arranged to suit piping layout requirements. A fractionation column or a large reactor has to be designed integrally with the associated pipelines, instrumentation, and access platforms. Often the vessel is heat-treated after welding, and it is essential to provide the necessary pads and connecting lugs for pipe and platform supports prior to this operation.

Having decided on the layout of piping relative to individual equipment items or groups of items, it is necessary to devise a system of pipe routing between equipment items and to the piperack (for which refer to the nine basic criteria listed above). Where possible, routes should be combined to allow for the use of common supports and a less cluttered plant area. As in the case of piperack work, it should usually be possible to arrive at acceptable configurations for hot pipes by the use of standard tables, and flexibility analysis should be a formal confirmation that the design is acceptable. Loads placed on equipment, especially on sensitive items such as rotating machinery, should be checked for acceptability.

19.3 Vessels

Unlike the connective items of steelwork and pipework, vessels are essentially equipment items, and are therefore of less significance in a text principally aimed at overall engineering rather than discipline specialization. Vessels may well be purchased in the same way as an equipment item, by the issue of a specification and a data sheet giving the required dimensional outlines and process conditions. This is particularly advantageous for pressure vessels, which are invariably subject to a design code which embraces both design and manufacture. These are interrelated, and it is not a good idea to split the final design and manufacturing responsibility. By 'final design' is meant the detail design of a vessel for which the purchaser has specified the following.

- The service conditions for which the vessel must be suitable. These inevitably include 'normal' and 'design' figures, to allow for process fluctuations. The 'design' pressure allows for the operation of safety devices in a way that is prescribed by the code.
- The materials of construction and corrosion allowance. Normally the purchaser (and ultimately the party responsible for process technology) is expected to be more knowledgeable than the vessel vendor about how to combat corrosion in the service environment, and takes responsibility for this aspect. This can also be achieved by the specification of generic types of materials which are acceptable (for example low-carbon steel), leaving the vendor to decide within this envelope on the most cost-effective solution for the service conditions stated, subject to the final approval of the purchaser.
- The functional requirements of the vessel: capacity, general shape and dimensions, internals, type and overall dimensions of support, piping and instrumentation connections and access openings, lifting lugs, and positions of pads and brackets for miscellaneous steelwork connections.
- The code(s) of construction or list of acceptable codes.

- Supplementary specification (over which any conflicting code requirements take precedence) covering such items as material and component standardization, corrosion protection, standard of finish, inspection, and quality records.
- Reference to the statutory requirements in the country of service.

Large tanks are usually purchased in the same way, for example subject to API Standard 650. Small atmospheric vessels may be treated similarly, but may also be built without reference to any code if in a non-critical service (for instance water).

Below a certain size of vessel, it becomes apparent that the design effort is excessive in relation to the overall cost of the vessel, more so when the need for individual foundation designs is considered. It becomes economical to standardize, and it is reasonable to expect that a number of standard designs should be available for most applications. This is often particularly important for fibre-reinforced plastic vessels, where special tooling may be obviated. Often this is inadvertently thwarted by the process designers, who size vessels by standard formulae, usually based on residence time. This can be remedied by communication and consultation at an early stage of design.

Repeating and developing some of the remarks made in previous chapters, it is essential that the design of vessels be the subject of inter-disciplinary input and review, including the following.

- *Process*. Function: fulfils process and process control needs; minimizes corrosion and wear; ergonomically suitable for operators. Safety: adequacy of vents or relief devices; adequate containment of harmful fluids.
- *Piping*. Suits piping layout; nozzles correctly dimensioned; allowable nozzle loadings co-ordinated with loads from piping; pipe support brackets provided where needed.
- *Instrumentation*. Correctly positioned and accessible connections, dip pipes, etc. where needed; clear understanding on limits of vessel/instrumentation interface and supply responsibility.
- *Mechanical*. Manholes, isolation facilities, vents, drains, cleanout/steamout facilities, and access all acceptable for entry and maintenance; correct interface for agitators, etc.; correct shape and finish for internal lining system (if any); correct configuration and types of clip for refractory and insulation (for all vessels requiring them, but especially for heat-treated vessels); provision for in-line inspection (tell-tales, underfloor drainage) when needed; transport

and erection procedure properly thought out; suitable lifting lugs included; partial shipment and site assembly specified if needed.

- *Structural.* Ladders, platforms, and agitator supports correctly designed; clear understanding of vessel/structural supply interface; steelwork connection pads and brackets included where needed.
- *Civil foundations* (or supporting steelwork). Conform to code requirements or code advice (for instance, consult API 650 for tanks); position and size of holding-down bolts (are they really needed?); provision for thermal expansion between vessel and support, if needed.
- *Electrical.* Position and detail of earthing boss: try to find a way to standardize on this, without having the electrical engineer review every drawing!

All of the above are equally applicable to fibre-reinforced and other plastic vessels, which in particular are best ordered as equipment items, with final design by the supplier. Plastic vessels are in general less robust and shorter-lived than steel vessels. Care needs to be taken to:

- provide for ease of removal and replacement of the complete vessel in a maintenance context without cutting too many cables, platforms, pipes, etc.;
- ensure adequate reinforcement and/or gussets for all external-load-bearing or potentially breakable members, for example pipe nozzles (especially small ones);
- ensure that the vessel base or support details meet code requirements, and are agreed with the vessel supplier; and
- provide adequate fire protection and retardant in the resins.

19.4 Chutes, bins, and hoppers

The layout of chutes, in other words their functional design, has been touched on in Chapter 16. This is a specialist subject which will not be developed further; a design office seriously engaged in this type of work should be expected to have a library of chute designs for various applications to cater for variations of geometry (including material trajectory), capacity, material characteristics, and lump size. Except for mildly abrasive applications, the chute layout has to correspond to a system of liners covering all abraded surfaces or lips (in the case of dead-boxes). The following design standards and preliminary work should be in place before proceeding with detailed chute design.

- The liner system, including standard shapes where applicable, the liner attachment system and corresponding seals, and an understanding of access requirements (if any) for liner replacement, which should be considered together with the access requirements for removal of blockage.
- The design requirements for dust containment and extraction, where applicable, including proprietary or other standard designs for dust containment skirts and skirting boards. (These should not be used to contain the main flow of material).
- For conveyor head chutes, studies of: the material trajectory and flow; the belt cleaner installation; access requirements for its adjustment and replacement; and the method of collecting dribble (and washwater, if used).
- The scheme for the breakdown of chutework for installation and removal and, in association with that, its supports.
- For chutes from bins, the design of spile-bar or other cut-off devices, if required.

Care should be taken to ensure that chutes handling large rocks are strong enough by comparing the proposed design with successful installations, on the basis that the impact energy is proportional to the drop-height and to the cube of the maximum lump size. For mining applications it seems to be inevitable that whatever the upstream size-grading equipment, the miners will occasionally find a way of loading conveyors with the maximum lump size that can be conveyed. This is probably about half the belt width of the conveyor, and it is wise to design for this value.

There is nothing to be added about the designs of bins and hoppers, except to reiterate the remarks made in Chapter 16 on design for materials flow, and to caution that for large bins or silos special consideration must be given to structural loads arising from mass-flow reactions (particularly in the region of flow convergence) and from uneven material distribution.

19.5 Civil design

The major components of civil design are:

- foundations and associated plinths and equipment bases;
- buildings;
- drainage;

- containment structures for process and utility substances (fluid and solid); concrete and masonry stacks and ducts for exhaust gases may also be considered under this heading when needed;
- concrete floor-slabs and associated bunds and drainage sumps;
- general site infrastructure, including roads, fences, and civil works such as dams, settlers, and evaporation ponds, when needed.

Some of the needs of civil design have to be anticipated before finalizing the plant layout, in particular the terracing and drainage systems, as discussed in previous chapters. It is also necessary to ensure that the plant configuration is not so confined that there remains inadequate space for foundations, whose footage area often exceeds that of the supported item, and for underground services. The architectural design of buildings is usually one of the final issues of layout development once the functional needs have been defined, and it is inevitable that architectural considerations may lead to review of the functional design and some iteration.

We will concentrate in the following on the design of foundations and equipment support blocks, since these are the civil items which are the most closely integrated with the overall plant design. Foundation design follows at its upper end from the plant layout and the loads and details of supported equipment and steelwork, and at the lower end from the design criteria produced by the geophysical consultant.

For proprietary equipment, a concrete base can be designed either by the project design team to suit the 'certified' vendor drawings of the equipment or, preferably, by adding details to the 'foundation outlines' supplied by the vendor as part of his supply obligations. In fact, the more complex the equipment, the more desirable it is for the vendor to participate in the foundation design or to supply the design. The foundations have not only to support loads, but also to maintain relative position of the supported equipment. The foundation designer must understand what movements can be accommodated, and design accordingly. For the more complex structures, it is often necessary to integrate foundation design with the detailed design of the equipment supported. The objective is to obtain the best compromise for a foundation system which can be expected to sustain limited deflections in the short and long term, and supported structures and equipment which can accommodate the predicted movement. Sometimes the solution to the compromise may involve the use of systems for ongoing measurement of foundation movement and for compensation by jacking and similar devices.

Support base design must also facilitate the initial positioning of the supported equipment, which may be problematic if for instance cast-in foundation bolts are employed. The designer should ensure that the system of dimensional tolerancing matches the needs of the supported equipment. For example, if two related machinery items have a small specified tolerance on their relative positions, the feature of the foundation design that limits the machinery positions, say the spacing of the cast-in foundation bolts, should be directly dimensioned one to another. If the relative positions are determined by dimensioning each set of bolts from an individual machine centreline, the possible cumulative tolerance error is tripled.

Where the dimensional tolerances of support bases appear to require the use of special construction methods or templates, it does no harm to call for that on the drawing. If the contractor comes up with a better method it will surely be agreed on site, while if no special notes are made the contractor may simply fail to meet tolerances which are considered to be over-demanding.

A design practice is necessary for foundations subject to vibrating loads. Usually the concern is simply to ensure that natural frequencies of the supported system are well away from the operating frequencies. Various design packages are available for such checks, but the inherent variability of soil properties (initially, as well as over time) detracts from the precision that can be offered by analysis. For many cases it is possible to rely on design practices by which rocking motion will be minimized, in particular to design foundations and bases which have a low ratio of height to width in the plane of excitation. For more critical applications, the involvement of the equipment vendor should be secured as part of his contractual responsibility.

Transmission of vibration to other structures can be a problem. Even on relatively simple applications, such as centrifugal pump bases, it is necessary to minimize the transmission of vibrations (usually by providing separate bases), otherwise standby machinery may be subject to bearing damage. In the case of control rooms, any noticeable transmitted vibrations are likely to be cause for complaint. Even when environmental standards concerning permissible control room vibration levels exist, and may seem attainable without special construction, it is advisable where possible to be conservative. Positively isolate the foundations and employ brick or concrete building methods if there is any vibration source in the vicinity, especially for low-frequency vibration exciters such as crushers and mills.

Most civil items within a process plant should be the subject of standardized design procedures and designs that require relatively minor development for specific applications – except for the designer looking for more work! Much of the actual design effort should go into ensuring the correct compatibility of the designs with equipment and structural interfaces, drainage, and access ways, and their correspondence to required plant layout and levels.

It is in the civil construction that relative positioning of the plant is largely determined. Cumulative tolerance errors should be avoided when addressing the plant as a whole. For large or complex plants, the person responsible for overall construction survey and setout should have the opportunity to discuss critical dimensions with the plant designer; this may result in a more appropriate system of overall dimensions.

The following is a list of some of the standard designs or design procedures which a process plant civil engineering team should possess.

- Bases for columns, pumps, vessels, and miscellaneous plinths.
- Bases for tanks: refer where applicable to the requirements of the standard to which the tank was constructed, especially for large steel tanks and all plastic vessels.
- Floor-slabs, including associated kerb, bund, drainage, and sump details.
- Expansion joint details to suit various environments.
- Drainage and sewer system details suitable for water and process fluids, including flammable and potentially explosive environments, and toxic substances where applicable.
- Acid-proofing details.
- Roads, kerbs, and associated drainage details.
- Electrical substations and control rooms.
- Standard architectural details.

19.6 Instrumentation and control

Detailed instrumentation and control design may be grouped in the following series of activities.

- Process control systems and the corresponding logic diagrams, software, and mimics.
- Field instrumentation and control device specification and selection, and the corresponding loop and schematic diagrams.
- Wiring diagrams, input/output diagrams, and interconnection

diagrams: whatever is required to indicate the physical interconnections of the component parts for construction.

- Instrumentation field installation drawings; control room layouts and the associated hardware configuration; panel construction drawings; cable and cablerack routing drawings; and the associated material take-off and cable schedules. In other words, the realization of the diagrams in space.

Process control system hardware and software is clearly a specialist subject which continues to evolve at a rapid rate. A control system comprises an information processing system, an operator interface, and a process interface. The information processing system may output process information to which the operator responds via a control instruction, or it may be configured to respond directly to sensed information, including cascades of directly interfacing higher-and lower-level information processors, not necessarily on the same site. The response may be programmed according to complex relationships and calculations, with the possibility of embellishments such as self-optimization via a neural network system. For the present purposes, the information processing system can be regarded as just another piece of proprietary process equipment, which has to be specified and purchased. The specified performance must include the following.

- The description and numbering of field devices (instruments, controllers, and sub-processors) and the corresponding I/O (inputs and outputs).
- The operator interface. The principal interface is usually a mimic displayed on a video screen, and the associated programming is performed together with the information processing program. Other interfaces may include data-loggers, audible alarms, and a computer station for program alterations, fault-finding, and general system maintenance.
- The logic by which the inputs and outputs are to be related (or for initial purchasing requirement, the processing and programming capability).
- The required flexibility for future modification and expansion; a substantial reserve I/O and processing capacity are usually needed eventually.

The task of preparing logic diagrams, specifying and selecting the processors and associated hardware, and of programming, calls for much inter-discipline involvement. This must include the process engineers ultimately responsible for designing an operable plant, the mechanical engineers procuring equipment packages which incorporate various instruments and controls and the electrical engineers configuring the switchgear, motor controls, and miscellaneous power supplies. Ultimately, there is also inevitably a heavy commissioning involvement, in program installation, debugging, and optimization. In resource planning for this element of project work, it is advisable to ensure the continuity of personnel from the software design team through to commissioning.

One other aspect of the work of overall control system design is the need to ensure stability. Possible causes of instability may already be seen from the P&I diagrams, if individual control loops separately control interacting process functions. Further instabilities may be introduced by the control logic. There may also be areas in which the need of experimental optimization of the control system may be foreseen. To address these concerns, it is necessary to look at the control system as a whole, rather than as strings of individual cause and effect. For complex control systems, to minimize the commissioning duration it may be worthwhile to carry out a simulation. This can be carried out on at least two levels: by building a computerized dynamic model, and by shop-testing the completed control hardware system, fully programmed, in much the same way as carrying out shop tests on a piece of equipment.

Field instrumentation design is a more self-contained and standardized activity, aside from the physical plant interface, which we discussed in the previous chapter. The input information from P&I diagrams and individual instrument data sheets requires little interaction other than the need to challenge whether the accuracies and specified details are really necessary, considering the costs and other consequences of which the specifying process engineer may not be aware.

Most instruments and control devices are of a standardized off-the-shelf design, which complies with one or more national standards. In certain cases, for instance flow measurement orifice plates or control valves, application calculations are necessary, but these are in general in a standardized format. Loop diagrams also tend to follow various standard formats, according to application. In summary, most of this work can be handled in a systematic way, leaning heavily on databanks of item design data and application software.

The same remark applies to the remainder of the instrumentation and control work. Wiring and interconnection diagrams can be produced in a fairly mechanized fashion once the connecting hardware details are known and a system of panels and marshalling boxes has been set up, appropriate for the plant layout. The individual instrument installation details should mostly be in a standardized format corresponding to instrument type. The actual physical plant interface has to be carefully co-ordinated with the interfacing disciplines, in some cases at the conceptual design stage, as mentioned in the previous chapter; but once this has been done there is seldom any need for iteration. The cable and cablerack runs can usually be designed (without much need for iteration) after the piping and electrical rack designs have been finalized, because the space requirements are comparatively modest and positioning requirements are comparatively flexible.

19.7 Electrical design

The work structure of electrical design is similar to that of instrumentation, the tiers being single-line diagrams, schematics, wiring diagrams, and cable/cablerack routings, together with associated schedules of quantities. There are also various electrical field devices, such as pull-wire switches on conveyors and electrical actuators, which from the design management perspective, can be treated much the same as field instruments (see Fig. 19.2).

Electrical switchgear panels are invariably detail-designed by switchgear vendors; a single-line diagram, specification, and schematic diagram for each motor or consumer circuit (or typical diagram for each type of motor/consumer) are sufficient information for purchase. An important (and sometimes schedule-critical) part of electrical design work is the layout, for further development by the civil designers, of motor control centre buildings and electrical substations generally.

A competent design office should have available standard designs and design systems for these buildings, incorporating appropriate spacing and floor cutouts to accommodate vendor-designed panels. Planning of equipment purchase and building design should be structured to eliminate possible delays to the buildings' construction while small details of switchgear design are finalized.

Electrical rooms and transformer yards require special fire protection, which should also be the subject of standard design systems, although usually sub-contracted to specialist vendors. The layout of cables, cabletrenches, and cableracks needs no particular elaboration other than the obvious need for close co-ordination with other disciplines to eliminate clashes and preserve access.

Fig. 19.2 Piperack/cablerack junction with valve access station – requires forethought and planning

19.8 Other technical specialities and consultants

Most project design teams include the disciplines addressed above, but it is seldom possible or economic to cover the full field of design expertise required for every project without the assistance of some consultants, design sub-contractors, or part-time specialists. The possible need for such assistance should be considered at the engineering planning stage, as part of the overall resourcing plan.

When using specialists, it is desirable to obtain the maximum benefit by getting the specialist's input at the earliest possible stage, before design criteria, layouts, and work-plans have been finalized. There needs to be some flexibility to agree the scope of the specialist's work, rather than dictating it, otherwise it is possible to employ a specialist merely to be told that the overall plant design, as configured, is not suitable.

It is of course possible to take the desire for specialist input too far, and end up with a project team who refer every important decision to an external consultant and take rather little responsibility themselves. Not surprisingly, such views may be strongly supported by the specialists! Many consultants have well-developed techniques for expanding their work on a project, usually by generating uncertainty either in the project team or through to the ultimate client and outside environment. Certainly, no consultant has any interest in abbreviating his work, and possibly putting his reputation at risk, while being paid less. The project engineer has an inevitable duty to resist such pressures and achieve a realistic balance. As for so many other issues, this is facilitated by planning ahead, by realizing the potentially manipulative nature of the relationship, and by ensuring that before the consultant is engaged he has made an informed commitment that includes a definition of responsibility, a scope of work, and a budget.

The following is a checklist of some of the specialties that may be needed.

- Geophysical consultants, to conduct site soil surveys and make appropriate recommendations on foundation treatment. (Note: do not allow these consultants to get away with a discussion on alternative possibilities; insist on explicit, quantified recommendations).
- Refractory design engineers. For many plants employing thermal processes, the performance of refractory linings is the most critical factor determining plant reliability and the need for shutting the plant down for repair. Refractory design work often includes the testing and inspection of raw materials, which tend to have variable

properties depending on their source, and may react with the process materials. Special attention may be needed to follow up from the design to the refractory installation, which is an art rather than a science.

- Heating, ventilation, and air conditioning.
- The design of systems for fire prevention, detection, and extinguishing.
- Corrosion, and corresponding metallurgical and corrosion-protection issues.
- Noise control.
- Hazard analysis and mitigation.
- Statutory compliance.
- Pressure vessels – possibly a code requirement.

In general these needs should become apparent from an appraisal of the critical aspects of plant design, and of the resources available.

19.9 Overview of the design process

A design operation consists of a number of decisions on how something is to be configured. These decisions are taken by a designer who has a certain 'window' on the relationship between what he is designing and the plant as a whole. His window is likely to be influenced by how his work is grouped. For instance, if all he does is design chutes, although he may become a very good chute designer, he may not be well-positioned to develop chute designs that take into full account the implications of different plant environments. He may design a functionally excellent chute, but without any thought to the load transmitted to adjacent structures, the best way of configuring chute supports for load compatibility and ease of erection, the access for liner replacement, and so on. Design review and checking therefore has to be conducted through a sufficient number of windows to include all possible consequences to acceptability.

Considering the most general form of checklist, each item design has to be correct or acceptable in a number of regards.

- The item has to fulfil an individual function efficiently, safely, reliably, conveniently for operators, and for an acceptable duration. For these purposes, the item design can normally be configured by reference to plant performance requirements, previous successful designs, and the application of design codes and standards.
  Example: *A pump*. The functional requirements are normally set out

on a sheet of data which includes the performance and the design code, for instance API Standard 610.

- The item has to interface to the rest of the plant (a) functionally, (b) by structural loading, and (c) by orderly occupation of plant-space.
  Example:
  (a) The pump has to be mounted, relative to the liquid source vessel, such that suction is maintained, NPSH is acceptable, and the suction line is reasonably short.
  (b) The pump support must be adequate for its service mass, and the pump nozzles must match the loads imposed by pipe connections.
  (c) The pump must be accessible for operation and maintenance.
- The item design must be acceptable and economic for manufacture, erection, and maintenance purposes, which may involve the need for standardization. (This is addressed in the specification and selection of the pump; see Chapter 13.)
- The design has to be represented in a way that facilitates its manufacture, erection, and maintenance, according to a system of order that best suits the overall project. (The pump data sheet and specification are configured according to API 610, a standard with which the target suppliers are familiar.)
- The item design may have to comply with other requirements specified by the client or locally applicable regulations.
  Example: Use of cast-iron pumps may be prohibited for flammable fluids, due to risk of casing failure when extinguishing a fire.

The design process is facilitated, and its accuracy and economy are improved, by utilizing standardized and previous project designs and design systems as a base as much as possible. The existence and availability of such designs and systems/checklists are measures of the acceptability of a design office engaged in this work, and the minimum requirements mentioned above may be used to check this. Design co-ordination and interface suitability are as important as the design functionality of individual items.

19.10 General arrangement drawings and models

When outlining the simplified sequence of engineering work in Fig. 4.1, we showed the production of general arrangement (GA) drawings as an intermediate step between the layout development and the production of drawings for pipework, platework, etc. The general arrangement drawings can in fact be used in three ways:

- as a means of establishing precise interfacing dimensions against which the piping and other items will match (before their detail design);
- as a means of verifying the match and the conformity to spatial requirements of the detail designs, drawn up directly (without the intermediate GA stage) from the layout drawings, as were the 'sketch' designs of Fig. 4.2; or
- as a combination of the previous two, generally in parallel with detail design development.

General arrangement drawings are essentially a scale model which facilitates design co-ordination, although they also aid construction. Physical or virtual models fulfil exactly the same purpose. However, it is possible to build a plant without such drawings or models: especially with computerized or other design overlay techniques, skilled draughtsmen may pride themselves on their ability to do this without interface error, and this is one of the techniques commonly employed for engineering schedule compression (discussed in Chapter 27). It is invariably found that this practice carries with it a cost in terms of design error, in particular at the interfaces, and that the frequency of error rapidly escalates to an unacceptable level for larger or more complex plant, which has more interfaces.

The strictly logical procedure of drawing the general arrangements before the details is the same as the technique of developing a complete computer model and then deriving all the individual details from the model database. The two have the same limitations.

- The detail drawings cannot in general begin to be released for manufacture until each reasonably self-contained section of the model is complete.
- The larger the individual sections of the model, the greater the delay in completing the model. (It is not efficient to have too many resources inputting simultaneously to the same section.) But the smaller the sections, the greater the number of interfaces created and problems arising, *unless the section boundaries can be and are carefully selected to minimize these problems.*

In conclusion, there is quite a complex system of information flow involved in design co-ordination. Its mastery for large and complex projects is a matter of developed teamwork, often supported by advanced software systems, but in any event greatly facilitated by careful planning and work breakdown.

Chapter 20

Traditional Documentation Control

A 'document' is the name given to any piece of recorded information, be it a drawing or a computer file. From the preceding chapters on project and engineering management, the paramount importance of documentation control for the project in general, and for the engineering work in particular, must be apparent. The product of engineering is information. Project management is the orderly and logical flow of information. Destroy the system of order within which information is presented, and it becomes worthless.

The central requirement of documentation control is a filing system, which has to provide the following functions.

- Document registration and indexing.
- Record and control of document revision (for instance Rev A, B, C...) and status (for example, preliminary, approved for purchase, approved for construction, etc.).
- Issue and receipt of documents, and corresponding record and control. Circulation of documents for comment and approval, and their tracking during this process, may be included under this heading or as a separate function; either way it is a labour-intensive but essential operation.
- Safe-keeping and archiving of documents.

The amount of work needed to do this properly is too often underestimated, with consequential overload and inefficiency. The document control system is the nervous system of the project, overseeing the transmission of information. Both order and speed are essential. This means that the resources available for this operation must *exceed* the average need, or there will be delays during peak loads.

It usually pays handsome dividends to employ at least one technically knowledgeable person in this function, thereby promoting quick action in the case of documentation, and especially vendor documentation, which is unacceptable due to inadequate content or incorrect format.

The nature of the work is repetitive and easily codified in a procedure. The number of documents to be processed is predictable,[1] and the timing is more or less predictable. The work can and must be planned, and adequate resources made available, on pain of losing control of the project when, for instance, it takes a week to circulate a batch of vendor drawings because they must first be renumbered in the project system, but the document controller is overloaded. And the delay may only be recognized when the time has been irretrievably lost.

Document indexes are usually based on classification by type of document, serial number, date, and revision. They can also be classified by subject, size, source, and sometimes by destination. There are also very many cross-references which can be useful: purchase order number, responsible project co-ordinator, etc. There are many possible schemes for numbering systems.

As we have already remarked under the heading of planning of engineering work, the work is planned and controlled by identifying each item of work by the associated document; hence the document numbering system should be compatible with the system for engineering work classification. Apart from documents, there are all sorts of other numbers with which work is planned: equipment numbers, plant areas, purchase order numbers, contract numbers, and so on. There are obvious advantages for a document numbering system which is compatible with other numbering systems, which together address the needs of the project as a whole.

The document and other item numbers may have to conform to client requirements for project tracking and audit, and much of the documentation and its indexing remain important for plant operation, maintenance, and financial control. The client use of project documentation is an inevitable aspect of any project numbering system. Even if the solution adopted is merely a duplicate set of numbers for post-project use, this can consume a lot of effort; this is minimized if the duplicate numbering is planned from the outset.

Issue and usage of a document, and change control, are all facilitated

---

[1] Predictable, with experience. The number of internally generated documents should include a growth provision (mainly for extra small purchase orders and contracts), but the major underestimation is usually in vendor data (drawings, manuals, etc.); if in doubt, refer to similar previous work to estimate these.

and indexed by reference to the document number. Nothing can cause greater confusion than a change to the document number itself; in fact usually the only way to handle this is by cancelling the document and creating another.

Aside from the filing-system aspects of document control, there is a need for standardization on the contents of documents in order to make the information contained accessible, and to make it possible to know what information is to be found within which document: an information filing system within individual documents.

In conclusion then, at the outset of a project it is essential to have a well-thought-out document control procedure and numbering system that meets the needs of all users before and after project completion, and obviates any subsequent need for renumbering.

Fifth Cycle

More on Management

Chapter 21

The Organization of Work

21.1 Packaging work

The organization of project work is one of the principal issues of this book. In Chapter 7 we discussed various ways in which a client could set up a project, and the implications of each choice, but we did not go very far with the organizational breakdown of the project work other than to recognize that the need to create competitive inputs may result in the fragmentation of the work. We have for the most part discussed the functions and work performed by the project team and their technical co-ordination rather than the different ways in which the project work components can be organized as a business.

Several conclusions relating to work organization may have been drawn.

- Smaller plants are much easier to design (smaller in the context of number of components rather than physical size or capacity), accurately and quickly, than larger plants, evidently because there are fewer internal interfaces, and therefore design iterations and interface errors are reduced. Construction organization may be simplified and site interfaces reduced, bringing benefits such as reduced interference between different contractors and more competition between smaller, more focused construction organizations.
- Breaking larger plants down into packages, configured such that each package has the minimum number of external interfaces, should therefore also offer the benefits of smaller plants.
- Better-value designs and project performance are promoted by specifying and paying for required output performance, rather than inputs or any intermediate functions or activities which are not part of the end product. Work paid by the hour always consumes more hours than work paid by the product item, and there are plenty of

hours of the day or weekend into which to expand the work. Paying for work by the month, and then paying overtime or other benefits for overtime, is the same as paying by the hour.
- Performance of all activities is enhanced by competition, provided that workscope is clearly defined and product quality is adequately maintained by setting acceptability criteria and by inspection.
- Working efficiency is improved, and errors are reduced, by adopting standardized designs and working procedures.

These observations are surely what common sense would predict, and yet in many instances plant owners and project managers continue to believe that value may be obtained otherwise. For instance, they may blindly put their faith in professional people paid by the hour, and be surprised when the work difficulty increases and it has to be performed at a rate of 70 h per week or more. Or lump-sum contracts may be awarded on a highly competitive basis, without in-depth surveillance and inspection of the work, with the result that all the gains of low price are lost in poor performance and poor quality. The delivery of inherently poor quality, even if not discovered until after payment has been made, may seem to be a simple problem to rectify legally, but in practice it is usually accompanied by incessant argument and obfuscation. Quite often, a poor-quality supplier or contractor concentrates his attention on surviving in the relationship long enough until any move to transfer the work to a competitor would be intolerably disruptive to the project. He then utilizes this position to negotiate better terms, failing which work performance will further deteriorate or stop. Here is a dilemma for the project manager: how does he get the best prices for his DFC components, without having to spend, what was thereby saved on increased surveillance, inspection, and contractor substitution costs, and increasing the IFC cost component? He must find the best balance, which will not be the same in different circumstances.

In the light of the above, let us now examine some of the issues which may arise from the way in which the performance of work is organized for major projects. The classic system of organization is based on the employment of an overall 'owner's engineer', which/who may be part of the owner's organization, or a consultant or possibly an engineering company. This is the contractual engineer who is charged with:

- the conceptual design and control estimate;
- the definition of packages of work, which together fulfil the project objectives (or of a single turnkey package);

- supervision of the process of contract award; and
- integration and control of the contractors' work.

This is of course similar in principle to what any engineer ends up doing at a lower level of a project. The important difference lies in the relationship with the owner, which has to be to some degree one of trust, in that the work to be done by the engineer (as for the project itself) is not yet defined. Thus the price for the work cannot realistically be set by a competitive process. The world of consulting engineers has many formulae to deal with this situation, but the basic alternatives are:

- payment by the hour, with the difficulty of adequately defining what an hour's worth of input should constitute and achieve, and the inevitable pressure to expand the number of hours; and
- payment of a percentage of the constructed cost of the project, leading to the absurd situation in which the engineer is expected to diligently pursue project economies in order to be rewarded by a reduced fee. (By inference, this work structure can be expected to lose potential economies of the DFC component.)

Evidently then, a realistic plant owner will seek to motivate his consultant[1] by means other than cash (although the cash component is of course essential), in particular by the benefit of an ongoing srelationship in which recognition is given to the established value of past performance. And recognizing that this type of relationship is not the most effective for getting the best results economically, the scope of work of the consultant should be limited to the minimum, which is to conceptualize, define, integrate, and control the minimum number of packages – preferably only one, if there is adequate competition on that basis, thus limiting the work of integration.

What is true for the project as a whole should be true for individual components of the project – and it is – but there are other factors to be taken into account. Firstly, at the lower level the component packages should be better defined, and a budget (or performance target) should already exist for each in order for a budget to have been set for the whole. Secondly, as the number of package components increases – which it has to, working down the organizational pyramid to eventually cover all the engineers, designers, and activity managers required for the whole plant – the quality of communication between groups working in

[1] 'Consultant' in this context may be an in-house role: whoever has the technical ability to, and is charged with, appointing one or more contractors to carry out the project work, or indeed, setting up an in-house team for the purpose.

parallel becomes increasingly more critical. Rather than studying the increasing influence of these factors as we address lower levels of the organizational tree, let us proceed to the lowest levels – the groups who actually produce the drawings, procurement information, and other outputs which directly configure the project hardware.

If a designer (or a group of designers) is working on a subject where there is a constant interchange of information with others outside his control, it is not realistic to hold him solely responsible for his efficiency. Certainly, performance indicators such as hours and errors per drawing can and should be trended, and some general conclusions on relative efficiency deduced, but it will have to be accepted that there will be many explainable performance deviations, and that is how most design offices control their work. It is possible to work on the basis of a lump-sum price per drawing, or group of drawings and/or other tasks, but only if virtually all the relevant information can be assembled before starting the task. This is usually very difficult, due to time-schedule constraints and iterations necessary for co-ordination with other disciplines. Where it is decided to work in this way, reference should be made to Chapter 19, Detail Design and Drafting, which should assist in determining which parts of detail design work can be split away as packages, and which parts have too much conceptual development content or too many interfaces to make a lump sum realistic.

On large enough projects, between the organizational level of groups of single-discipline drawings and the level of overall project management, it should be possible to structure groups of activities, where the majority of information interchange is internal to the group. The group of activities can be handled as a single performance area – a project within a project – which is paid as a lump sum. The objective here must be to arrive at a relationship where both the employer and the employed have the opportunity to benefit from improved performance, without losing the benefits because of the increased efforts of managing the arrangement.

Grouping work-teams on the basis of individual process units (or sub-units) within the plant is clearly the most effective arrangement for the latter purpose. If these can be organized as individual turnkey packages, there is the added incentive that decisions in one discipline have to directly consider consequential costs in other disciplines, for example, the additional foundation costs incurred by a particular machine design reflect in full on the costs incurred for the package. This grouping also lends itself to a clear recognition of the technical acceptability of the overall design: if a plant unit performs poorly, or if it is inaccessible to operate or maintain, there should be little doubt as to where the responsibility lies.

However, there are disadvantages in trying to set up individually accountable work-groups; these include the following.

- The time and effort required to set up the arrangements.
- The relative inflexibility in manpower allocation and work transfer between groups.
- The myopic vision that must be expected (and compensated for by appropriate surveillance) when relatively narrow responsibilities are allocated.
- A lump-sum payment system is not suitable for cases when the organization has to be set up before the conceptual design is finalized.
- For projects where lack of work definition has compelled the client to pay for the overall project by the manhour, it is invariably impractical to develop any other arrangement for lower levels of work-group. That is to say, if a work-package manager is paid according to the hours of input by each member of his team, it becomes impractical for him to pay the team members themselves a lump-sum basis per output. It is unmanageable.

Many different work structures are common in practice, depending on the circumstances of the project and the culture of the work-force. Whatever is contemplated requires analysis of its advantages and disadvantages, and planning of how to compensate for the foreseen drawbacks. There follows a brief checklist of features to consider when analysing a given proposed structure and asking the question of how the structure deals with each issue. Many of these features impact on each other, but they still need individual consideration.

- Accountability:
  - Are the objectives and workscope of each work-group clearly defined?
  - Is performance (in terms of quality, DFC, IFC or own cost, and time schedule) of each work-group measurable and adequately reported?
  - Is there an adequate mechanism to hold each work-group accountable for the 'knock-on' effects of their work on other work-groups? Knock-on effects include increased DFC, increased IFC, delayed or incorrect or inadequate information, and disruption due to changes.

- Motivation:
  - Are the conditions of employment directed to motivate the work-groups to meet all the project objectives?
  - Are there any additional low-cost benefits which may significantly promote productivity and quality?
- Communication:
  - Are important interfaces connected by direct communication?
- Standardization:
  - Are work practices, documentation and numbering systems, and design features standardized as much as possible across the work-groups?
- Duplication and omission:
  - Are any functions duplicated?
  - Are any requirements of the overall project objectives and workscope omitted?
- Organizational hygiene:
  - Is the proposed overall administration system likely to cause any unnecessary frustrations?
- Management complexity:
  - Has there been a detailed evaluation of the full extent of overall management work and manhours to adequately manage the proposed work structure?
  - Is the cost and effort justified by the benefits of the proposed organization?

21.2 Procedures

'Procedures' is a surprisingly controversial subject, in that one encounters quite a few apparently competent people who are quite antagonistic to any form of written procedure, and use words like 'red tape' to describe them. And yet a procedure is just a way of doing something, no more and no less. As there are inevitably good, bad, and unacceptable (even dangerous) ways to perform project activities, developing and using sound procedures is an inevitable necessity. Furthermore, to work in a team where efficient communication and co-operation are essential, and where work is planned and controlled, procedures need to be uniform.

Undoubtedly the reason for widespread apathy and occasional antagonism on this subject is that, so often, procedures contain unnecessary verbiage and unnecessary activities, and generally over-complicate the work. And if this is the reason why work is often poorly or wrongly performed, when observance of available procedures would have improved the outcome, then it must be a major issue. Perhaps the problem may also be that the preparation of procedures to do a job properly is regarded as unproductive use of time; the old syndrome that there's never enough time to do a job properly, but always enough time to do it twice.

So the subject here is the preparation of user-friendly procedures. And just by using that last adjective, and noting its prime importance, perhaps some improvement is already possible. Here are a few pointers to assist in the production of user-friendly procedures.

- A procedure should be clearly divided into that which is mandatory or essential, and that which is optional, guidance, advice, example, and so on. The mandatory/essential should be kept as brief and concise as possible. No requirement should be mandatory unless it is essential. Nice-to-dos, guidance, etc. should be cross-referenced and relegated to non-mandatory appendices.
- Where possible, use standard well-thought-out forms, and simply refer in the procedural text to 'Complete form B'. It is advisable to format forms, when applicable, so that it is clear what is mandatory and what is optional.
- Make maximum use of workflow charts, which include in every activity box the person, who is responsible for the activity. Do not provide a commentary that duplicates the information in the workflow chart. If explanation is required, put it in a non-mandatory appendix.
- Every procedure should include an authority designation, which may agree to modifications to part or all of the procedures where circumstances require it. The authority level should be as low as the organization can tolerate, depending on the subject.

This last point touches on another facet of possible antagonism to procedures, which is that of ownership. When someone describes a procedure in terms of 'red tape', he is implying that it not his procedure, it does nothing for him in the pursuance of his prime objectives – it is a burden placed on him. This may be a very important motivational issue, especially for very demanding projects where the apparently impossible is demanded. The attitude of the individual may be 'I'm supposed to do the impossible, with one hand tied behind my back.' And there is another slant to this, which is that enforcement of unwanted procedures can breed an unproductive attitude, which does not help performance improvement: 'If that's what you want, that's what you get'.

Evidently, if they are working in a productive environment and initiative is expected, the project performers must have some leeway. That is not to say that all procedures should be optional: this has been tried often enough – it is risky on small projects and disastrous on large ones. The ideal is possibly to have two sets of procedures, the first of which is mandatory to the project team, and the second of which will be developed by team members who are as close to the level of performance as possible, subject to whatever approval seems sensible. It is important of course to ensure that individual accountability and measurability of performance are retained. The first set of procedures may include matters such as documentation control, financial accounting practice, progress and cost reporting systems, and requirements for engaging in contracts; namely those practices that make project work auditable and comprehensible to others. The second set may simply be a list of other procedures which are usually found to be necessary, backed up by examples found to be acceptable in the past.

Of course, all that has been mentioned in the last two paragraphs is simply an enlargement of what was stated in the fourth item above: there need to be authority levels, as low as possible, for approval of modification to procedures. Mandatory content decreases as authority level increases; the trick is to keep the mandatory element to the minimum at each level.

Procedures, like plant designs, need iterative review and improvement to reach an acceptable standard – when there is nothing left to take out.

Chapter 22

Construction

The following is a brief review of some of the more important construction management topics, with special emphasis on the engineering interface and the peculiarities of process plant work.

As for other project activities, there are two basic ways to procure site construction: either to have the work performed on a product basis, by which one or more parties contract to perform parcels of work at a price per item, or to have the work performed on an input basis. For the latter, the project manager hires the construction labour and supervisors on a time basis, either directly or through a contractor, and also hires or buys all the other resources of construction plant, tools, temporary facilities, and consumable materials. There are many pros and cons of each alternative, including the availability and competitiveness of potential contractors, and the need for flexibility in control of site activities, which may be necessary in the case of unpredictable site delivery dates for materials and a likelihood of late design changes.

The basic way in which construction work is organized obviously affects the way in which the various parties involved relate to each other, and to the way that workers are motivated. However, the essential work content and management needs remain the same. As for other project management work, there is plenty of scope for moving work content between the 'indirect' costs of project management and the 'direct' costs of contractors. In the following we will ignore the differences arising out of the way that the work performance is contracted, and instead concentrate on the essential technical and management activities that have to be performed.

22.1 Survey and setout

The first site-related activities are invariably pre-project requirements, including the identification of the site and verification of its suitability, or possibly the survey and comparison of several sites followed by the choice of the best site. The aspects of 'survey' to be addressed are all those factors which may affect the operation and construction of the plant, such as accessibility, contours, sub-surface conditions, existing structures and facilities, and availability of labour and utilities, as well as simply observing and recording the topographical co-ordinates.[1] Arising from these activities is a site plan, which is based on a system of co-ordinates related to physical beacons and benchmarks, used for setting out the site works and controlling spatial relationships in construction. Needless to say, the quality of the survey work by which these functions are performed must be impeccable if costly errors involving the lack of fit of adjoining structures are to be avoided. A corresponding degree of care should be taken in the specification for survey work, the format and content of survey reports, and the selection of the surveyor.

22.2 Site management

The essential elements of site management are the following.

- General management and administration, including financial control, personnel management for directly hired staff and labour, asset management, site security, and standard office administration functions.
- Management of site construction work in terms of progress control against schedule and resolution of all the associated co-ordination and planning problems; cost control and management of cost variations arising out of changes and contractors' claims; and quality control, including acceptance of completed work.
- Management of technical information required for construction and technical records arising out of construction, and the generation or acquisition of any additional technical information needed for construction (in other words, problem-solving).
- Management of materials used in construction, including all items sent to site for incorporation in the plant and surplus (or spare parts) to be handed over for plant operation.

---

[1] Factors such as environmental impact are considered here to be the responsibility of another specialist consultant.

- Contract management and administration for all contracts involving work on the site.
- Safety management, usually retained as a distinct function to provide focus, although it is an essential component of the work of each supervisor.

Additional management may be required for specialist functions such as labour relations, the control and maintenance of construction equipment, and the supervision of rigging and welding, or these may be part of the overhead structure of contractors. And as for any other aspect of project management, the theory of limiting returns is applicable in determining the optimum input.

In the following, we will focus on some of the activities that are directly related to project engineering, or that require anticipation during the engineering phase.

22.3 Technical information for construction

Technical information that is necessary for construction includes copies of drawings, specifications, purchase orders, contracts, shop inspection reports, non-conformance concessions, packing lists, equipment vendor instructions, general plant installation data, and briefs[2] from engineering to construction management. Equally important are the registers: equipment lists, pipeline lists, drawing registers, and file indexes, by which the information is accessed and its status verified.

The timely provision of adequate technical information to site is as important as the timely provision of construction materials, but it is too often neglected by comparison. The competent project engineer will plan the system of information at the beginning of the project, follow up to ensure an orderly flow of information, and check that it is available on site and is utilized as intended.

Information in general, and drawings in particular, are subject to revision. Documentation control procedures must therefore include a routine for withdrawing outdated revisions, and since this never seems to be done perfectly, a routine is needed for immediate update of the registers by which document users are obliged to check the revision status before using the document.

---

[2] This subject is discussed in Chapter 25, Communication.

Information systems inevitably require a two-way loop, incorporating the return of comments and questions on information which appears to be wrong or inadequate. Considering the real-time needs of construction, the structure and performance of this loop is usually critical. A few ways of improving performance are outlined here.

- By setting up a clear procedure for dealing with 'site queries' which facilitates their record, transmission, and prioritization, and by ensuring that adequate engineering staff are available to deal with them. This may be a major problem if the engineering budget is low and has been expended by the time that the bulk of site queries arise.
- By empowering site engineers to deal with certain queries directly, without recourse to the original design team. This is not favoured for quality management, as the continuity of technical responsibility is lost, and it is probable that site engineers may be unaware of some important technical factors. More appropriately, the site engineers may make quick decisions which are later reviewed by the original design engineers, who can initiate the rectification of any unacceptable decisions. Experience here shows that it may be necessary to restrain the design engineers from letting their egos get in the way of their critical faculties!
- By moving the leaders of the design team to site. This is the best but most expensive option and often it is not possible until too late in the project, owing to ongoing office engineering needs.

In practice, a combination of these methods is normally needed. Too often there is a tendency to under-resource the site queries resolution function, especially on fast-track projects where many design assumptions are made, and a high number of discrepancies have to be resolved on site (a fairly predictable outcome). There may also be an attitude by the plant owner or project manager that this concentration of resources 'should not be necessary' if the engineering work is performed properly in the first place, even though half the problems are usually generated by manufacturing or construction error. Whatever the cause, false economies should be avoided.

22.4 Site materials management

Although at first sight this may appear to be a purely construction management issue, in fact materials management cannot be adequately performed unless the materials control system has been adequately engineered; this is especially true for bulk materials.

When construction work is performed on a contract basis, there are several advantages in including the procurement and management of bulk materials within the contractor's responsibility. Firstly, the corresponding engineering and management workload is shifted from the indirect to the direct cost report, if this an issue; secondly, the contractor is more in control of his work; while thirdly, a whole field of potential claims by the contractor – that of 'waiting for materials' – disappears. There are also disadvantages, in particular that material cannot be ordered until the construction contract has been awarded and its management commenced.[3] Much flexibility is lost in the ability to move work between contractors in accordance with their performance, and it is more difficult to provide for reserves of material for design changes. Overall it should be more efficient in terms of manhours for the project engineers to order the material, as they should be better able to utilize single-entry software, by which procurement information is electronically generated from the design drawings.

Whatever the reason, it seems to be preferred on larger process plant projects for most bulk piping, electrical, and instrumentation materials (except for smaller off-the-shelf items) to be ordered by the project engineers, who need systems to make and update take-offs, generate procurement information, facilitate tracking, and exercise cost control. The site materials control system has to be a part of these systems or fully integrated with them, using common commodity codes, take-off data, and ordering information, and feeding back information on materials shortages and materials consumed for site modifications and losses. Thus it is normally necessary for the project engineering function to devise and maintain an adequate, overall materials management system that is structured to be usable by site storemen, that is, user-friendly and not over-elaborate.

22.5 Heavy lifts

Equipment or structures that require special lifting gear or clearances have to receive attention from the design criteria development stage. Heavy lifts need to have a methodology developed at the conceptual

[3] It is possible for the project engineers to order long-lead bulk items (particularly valves, which can be ordered early from P&ID information), and hand them over to the construction contractor, who must procure the balance of items. This is indeed sometimes practised, but it brings problems of its own, in particular a duplication of systems.

design stage, and the practicality of these plans needs to be verified periodically as the project develops and before making construction commitments. If large and expensive mobile cranes are involved, especially on remote sites, the planning and cost control of this element of the project become just as important as the design and procurement work for major pieces of plant equipment, and consequently need to receive similar focus. The engineering responsibilities must include an understanding of acceptable rigging methods, the provision of appropriate lifting lugs, and schemes for disassembly for transport. If it is possible to second an 'early starter' from the construction management team to assist the engineers with this work, it should benefit from specialized knowledge, focus, and improved communication.

22.6 The 'site and office' relationship

A rather obvious statement, but one from which we have to commence here, is that the final objective of the project is the construction and commissioning of the plant on site. Ultimately all parts of the plant have to be integrated and to be simultaneously functional. To achieve this objective requires a planned and controlled effort of bringing together design information, materials and equipment, and construction workers and their equipment. Most of the site work has to be carried out in a certain sequence, starting with site clearance, then earthworks, foundations, erection of structures, and installation of mechanical equipment, followed by the interconnecting pipework, cables, and auxiliary items. Although the preceding project activities of design and procurement have a sequential logic of their own, and can be planned accordingly, inevitably the overall plan has to be construction-orientated, effectively working backwards from the overall plant completion. The backwards-looking plan almost invariably results in changes to the engineering and procurement plan by prioritization of critical items.

Whatever the starting point of the development of the project schedule, it must co-ordinate with the construction plan, the completion of individual design activities, and the delivery to site of materials and equipment. This is a statement of the obvious. An equally obvious statement, which however sometimes seems to be overlooked, is that in a project of any significant size, some of the activities or products which precede construction will be completed late. It is therefore an integral and fundamental part of the construction management function to make whatever arrangements are necessary to get the project back on track – it is a part of the job. The inevitable outcome is that construction management bear the brunt of engineering or supply shortcomings, and this can make for a difficult and counter-productive relationship unless managed wisely.

This relationship can be especially taxing in the case of projects that are fast-tracked, with many design assumptions made and, in the case of new process technology, with many process changes. As we have seen in the discussion on engineering and management optimization, there is no such thing as a perfect job – something which is an uneconomic proposition for those areas of work not identified as critical. Some errors and imperfect design and manufacturing outputs are inevitable. Of course, poor work can also be a product of a poor engineering effort, which this text does not condone; however, there is often a tendency by construction staff to regard all errors or imperfections as evidence of the incompetence and carelessness of the 'office' staff. This can be exacerbated by the attitude of an unknowledgeable client who of course wants perfection, provided that it costs no more, and often develops (when their by-product surfaces on site) a peculiar amnesia about the compromises and changes made earlier in the project.

The management relationship between construction work and the preceding work must be structured to accommodate the necessary flexibility. This is not the case if, for instance, the construction management operates as a separate entity with self-contained incentives, and any necessary adjustments to the work-plan are regarded as extra work, requiring additional incentive.

The latter situation is all too commonly arrived at either by inappropriate selection of construction management personnel or by contractually separating engineering and construction management. The engineering, procurement, and construction management functions can only perform optimally as an interdependent partnership, and any client or project manager who permits any one of the three to assume a privileged position has only himself to blame for a substandard project.

The project engineer who finds himself in a one-sided relationship with construction management may regain the initiative by taking a similarly one-sided attitude to technical failures in construction, namely, by instigating over-stringent inspection (say of non-critical welding) and by being slow and uncompromising in the issue of concessions. As in the case of the work which precedes construction, there is no such thing as a perfect construction job. The facility to take such balancing action against construction management (or indeed the client) may be enhanced, if the engineer perceives in advance that he is working on a 'political' job, by developing suitable defensive strategies. Typical examples are to plant some minefields in the specifications and contracts, even in the issue of concessions and tolerancing of drawings, which will be likely to require his assistance on site. Another technique is to hold back on information that is explanatory but not strictly essential, and to bury critical information among a mass of irrelevant data, enabling the engineer to gain the upper hand by exposing site incompetence. Of course, these games do nothing whatever for the project overall, but in a situation of manipulation backed by a client or project manager with limited vision or a hidden agenda, the choice is to be a victim or a player.

In the case of dealing with a lump-sum construction contractor rather than a construction management team, the same considerations generally apply and indeed may be even stronger. The contractor has a clear incentive for financial gain by over-emphasizing the consequences of delayed or flawed materials and information. In addition to the forms of counter-attack mentioned above (for use when needed to attain a balanced relationship), a common practice is to induce the construction contractor to enter into commitments based on inadequate data but subject to the contractor's prior responsibility to 'inform himself fully', such that he has no recourse when the need for unforeseen extra work arises. However, here we are touching on issues which form a field of (mainly legal) expertise beyond the scope of this book.[4]

In conclusion, the essence of the above is that in real-life situations, successful competitors sooner or later need to defend their positions in a process of negotiation. Those who keep their eyes open, think ahead, and build appropriate strategies are at a considerable advantage.

---

[4] Over the ages many consulting engineers, in particular, have made an art form of such contract clauses, tested in court following legal dispute and found to be effective.

Chapter 23

Construction Contracts

23.1 Structuring contracts

The preparation and award of contracts, especially those for site work, is one of the most critical aspects which can affect the performance of a project. Arguably the best approach is to deal only with experienced contractors who are in business to stay, and have a record of sound performance rather than of claims and litigation. However, there will always be degrees of contractors' experience, graduations of their perceived ethics, pressures to accept lower prices from less desirable contractors, and situations where little choice is available. It is necessary to prepare for the worst, and professional drafters of contracts are expected to make provision to deal with rapacious and incompetent performers. Some of the following observations will follow a similar defensive line, which is not applicable to the happy case of a contractor who delivers a fair job of work for a fair payment, without whining and trying to improve on the agreed conditions.

The foundation of a construction or shop fabrication contract is inevitably a set of general contract conditions, couched in appropriate 'legalese', which has been developed and proved to stand up in court for typical areas of dispute. The essential input from the engineer into the contract documentation is the description of what work has to be done, and how the payment will be structured in relation to the work. (The latter is essential because it directly relates to the way in which work quantity schedules are prepared.) The engineer's input must be consistent with the wording and intent of the conditions of contract, with which he must therefore be familiar. Most organizations provide the services of a legal/commercial specialist or a professional quantity surveyor for putting the final document together and ensuring that there are no loopholes or ambiguities. However, this specialist is unlikely to have the engineer's knowledge of information that determines the strategies to be adopted in a given instance, on which we will focus.

To be meaningful, the description of work for a contract must encompass:

- the technical specifications defining the standard and acceptability of the work;
- the quantification of the work;
- the description of circumstances which may affect the performance of the work; and
- the time schedules within which the work may, and must, be done.

It is usual to precede these specifics with a general description of the work.

Many of the remarks made about technical specifications in Chapter 13, Specification, Selection, and Purchase, are applicable for construction contract specifications. In particular, standard national and international specifications should be utilized as much as possible in the interests of contractor familiarity and of benefiting from the experience that goes into such documents. In the case of the erection of proprietary-design equipment, the specification should logically include the vendor installation instructions, drawings, and details of disassembly for shipping although, remarkably, many construction contractors are willing to quote a firm price for the work without sight of the latter.[1] The observations previously made on the planning and costs of inspection are also applicable.

The quantification of the work may be regarded as the preserve of a separate profession, the quantity surveyor, but many practitioners in the process plant field operate quite satisfactorily without such assistance.

For civil works, most countries have a standard defining breakdown and nomenclature of work operations for the purpose of defining schedules of quantities and rates.

Steelwork and platework are easily handled on a tonnage basis, provided that there is adequate definition of how the tonnage will be measured (preferably from the approved drawings rather than the

---

[1] This and similar observations have often led the author to wonder whether the whole process of bidding for lump-sump or rates-based work of this type is not really a complete farce. Perhaps both parties know (but will never admit) that there are so many intangible and unquantifiable factors affecting work performance that the end price will be fixed by negotiation, on a basis of what is 'reasonable'. If so, the observations made in the previous discussion on relationships are reinforced.

weighbridge), and of how bolt-sets and painting will be handled. It is of course necessary to provide typical drawings indicating the type and complexity of the work, and there may be arguments if the drawings available at the time of bidding are significantly different from the final drawings. Such arguments should be anticipated at the time of issue of the differing drawings, and resolved before the work is done.

Piping, instrumentation and electrical installation, and mechanical equipment erection are best handled by tying the construction work quantification to the quantities of material and lists of equipment to be installed.

For all of these work quantity definitions, there are schools of thought which believe that greater and greater breakdown of work input definition leads to more accurate, and therefore more competitive, pricing because there 'should be less contingency for the unknown'. For instance, in the case of piping, instead of relying on rates determined by an averaging factor which the contractor has to determine from the drawings presented to him, the piping work schedules may be segregated into:

- quantities of piping on piperacks at various levels, on sleepers, and at various levels around process equipment, with rates to be offered for each situation;
- each and every item of work that goes into the job, such as cutting a pipe, a weld-preparation, erecting a pipe or a fitting into position, a carbon-steel weld, an alloy weld, an alloy socketweld, a hydro-test, etc. – with a rate for each activity, per size and schedule of pipe.

Each combination of the two classifications above can be the subject of a specific rate.[2] Needless to say, the complications of calculating and verifying prices and contractors' invoices on this basis can be enormous, and it becomes apparent that the optimum degree of price breakdown is a matter of limiting returns. And one of the concerns of engaging specialized quantity surveyors for the job is that the tendency may be to spend too much money on excessively itemized price structures.

Paying for more itemized breakdown of unit operations can also lead to direct cost increases. For example, a job for which scaffolding is paid as a separate item (even under a separate contract) invariably seems

---

[2] Or a multiplier based on standard ratios of artisan's time per operation: this simplifies the number of different rates to be determined, but not the surveyor's work in determining the quantities.

to require twice as much scaffolding for twice as long, as when the provision of scaffolding is included in the end-product rates.[3]

The 'description of circumstances which may affect the performance of the work' (p. 266, bullet point 3) includes:

- the conditions on site, traditionally handled by a site visit (but ambient factors such as weather extremes, other hazards, description of site facilities, and regulations for safety, security, and for environmental protection require specific record);
- obligations of co-operation with other contractors;
- reporting, communication, and general work inspection and acceptance requirements; and
- special site facilities, such as potable water, power, etc., and their limitations.

The above are general to all contractors on a site, and best handled by a standard construction document, drawn up with due regard to the general conditions of contract applicable. In addition, there are some items which are specific to each contract:

- restrictions on access to specific work-areas at specific times;
- the definition of technical information to be supplied to the contractor; and
- the definition of free-issue equipment and materials.

The time schedule for the contract, to be meaningful, has to include not only the dates by which various parts of the work will be completed but also the dates by which the contractor will receive any outstanding technical information, equipment, materials, and access. These become obligations on the part of the project managers. Little purpose is served by omitting these dates with the objective of creating a one-sided contract, other than the probability of eventual dispute. Rather, quote dates which there is a commitment to meet and, if there is any possibility of not meeting them, the actions expected from the contractor in that event. Be prepared to deal with the situation.

---

[3] The moral of this is that separate payments should not be made for *any* activities, tools, or incidentals which are not a part of the end product: pay for the end product, not the means of getting there. It may be necessary to ensure by specification and by supervision that no unacceptable short-cuts are taken as a result, for example, that scaffolding is adequate for safety. If scaffolding is needed in the same place by more than two contractors, let them strike a deal between themselves.

23.2 Structuring payment to contractors

The basic forms of contract to consider are lump-sum, remeasurable, and reimbursable. In reimbursable contracts, payment is based on the contractor's inputs, essentially independent from any measure of the work output. The inputs may be based on a measure of actual costs (always rather hard to determine in the case of overheads) or on contract rates per category of input, such as per manhour for various skills and per week for construction equipment, with perhaps a separate overhead percentage. Clearly there is not much incentive for efficiency here, although bonus payments based on performance can be offered (with substantial complications). It is difficult to choose the most cost-effective contractor, and to understand what unit of output will correspond to a unit of input and how much effort will be made to complete the work on schedule. This contract basis is therefore normally reserved for small works, where the contractor can be replaced if ineffective, or for when the nature of the project makes it difficult to define sufficiently accurately the work description at the time that the work must be contracted.

The opposite extreme is the lump-sum contract, when a single price is quoted for the complete job, inevitably with provision for penalties or damages on late completion. Clearly, such a contract basis requires a precise statement of the work description, including all the factors listed above. Such contracts are rare in process plant work, but have their particular application and effectiveness in the case of repeat or modular plants, where the complete design is available before construction is commenced. It becomes possible to do away with all management and engineering costs related to item breakdown, quantity take-off, and detailed work measurement.[4]

The most frequently employed form of contract is based on item prices for erection of equipment, and the structure of unit rates for fabrication and construction of bulk items. This provides flexibility for calculating contract price adjustments to cater for quantity changes, and for a cost breakdown which facilitates price negotiation in the event of other changes.

We will not enlarge further on the structuring of rates, but will restrict

[4] At least at the project management level. The individual contractor will still require, for his own management purposes, documents such as quantity schedules for ordering materials and controlling work. He will have to generate these documents, which would otherwise be prepared by the engineer. Failure to recognize this can cause both dispute and substandard work performance due to working without proper controls.

further discussion to their possible manipulation. It is customary to provide (in addition to the payment of rates for measurable items of work) for the payment of 'P&G' costs. This is the standard term for time-based costs which compensate for the contractor's site establishment, including site offices, construction equipment and some overhead functions, and the set-up and removal of these items. Construction contractors often try to include as much of their costs under P&G as possible on the basis that the period on site will be extended for causes beyond their control, and their claim for extension-of-time costs will be inflated accordingly. In defence against this practice, a realistic allowance for time-extension costs should be allowed when comparing bids, and P&G costs should be itemized as much as possible, to obviate paying for items (such as heavy equipment hire) out of context.

Individual item rates should be scrutinized for consistency before comparing contractors' offers. Experienced contractors are often able to spot, in a pricing schedule, items whose quantities are likely to increase (which they load) and those which should decrease (which they lighten), thereby securing a higher payment than the summation of their bid for comparative purposes would indicate. Items such as 'dayworks' rates need a particularly jaundiced eye.

23.3 Claims

The settlement of contractors' claims can be a nightmare which turns an apparently successfully completed project into a disaster. Claims settlement is one of the most frequent causes of major cost overrun. Even if the claims are disputed and eventually legally rejected, the length of time taken and the effort involved can leave, over an otherwise successful project, a cloud of uncertainty that remains until long after the project achievements have been forgotten.

There is not much defence against the institution of unreasonable claims, other than the avoidance of litigious contractors. But such claims are likely to be seen for what they are, and are unlikely to be the subject of reservations as to the project outcome.

Evidently the subject of claims management is an important one, worthy of specialized input which is beyond our present scope. For most purposes, project engineers concentrate their attention on rapidly and reasonably settling any problems which may otherwise escalate into a contractual dispute, and above all on eliminating the causes of such problems.

The main causes over which the engineer can exercise some control are the following.

- Unclear and possibly inadequate work description (for example, of any of the elements listed above). Work should be described inclusively, that is to say that all aspects not specifically excluded are included. Let the bidders qualify their offers if they are in doubt.
- Late provision to the contractor of information, materials, facilities, or access (including access to work which first has to be completed by others).
- Changes or faults in information and materials supplied.

We will not dwell further on the subject of work description, but rather concentrate on the factors of lateness and error. These are in practice quite closely linked: working under pressure of time results in increased error. Obviously, one way to minimize potential claims from this source is to delay construction work until engineering work is complete and all materials and equipment are available. This generally is not an acceptable alternative, because of both the extra cost of financing an extended project, and delays in product availability. On the contrary, it is standard practice to appoint a construction manager very early in the project, even at the outset. This person will want to get on with the job, and be almost guaranteed to create the maximum pressure to get contractors on site and commence construction work.

If a project deteriorates to the point where both the quality and the timing of information and materials sent to site are inadequate, it can and usually does deteriorate in exponential style as more resources are diverted to the solution of problems and less are available for the advancement of work which is already behind schedule. In this context, construction claims and costs are bound to soar owing to the combined effects of disruption and extended site establishment.

The consequences of allowing the provision of materials and information to site to get out of hand in this fashion are usually far more severe than the consequences of initiating a controlled schedule delay. Unfortunately this is not readily perceived early enough, possibly out of optimism, more probably from collective behaviour in a climate where the acknowledgement of bad news is regarded as defeatism.

There can be many causes of unforeseen schedule delays, ranging from lack of competence of the engineering and management team to supply and shipping problems and unrealistic schedules in the first place. Another frequent cause is the acceptance of too many changes in plan, especially of design changes after detailed design has commenced.

We will defer discussion of ways of controlling changes and recognizing the point of spiralling deterioration until Chapter 26. The point to be made now.is that the best way of minimizing claims for these reasons is usually to make managed schedule changes (not necessarily including the end-date) as soon as the problems can be foreseen, and if possible before making the corresponding contract commitment.

Chapter 24

Commissioning

Plant commissioning consists of the following activities.

- Checking that the plant construction is complete and complies with the documented design or acceptable (authorized and recorded) design changes.
- Carrying out preparatory activities such as cleaning, flushing, lubrication, testing of electrical circuits and instrumentation loops, and set-up of control software. These activities are generally known as pre-commissioning work.
- Energizing power systems, operational testing of plant equipment, calibration of instrumentation, testing of the control systems, and verification of the operation of all interlocks and other safety devices, without yet introducing process materials. These activities are usually described as 'cold commissioning'. In parallel, it is usually necessary to commission the plant utilities, such as cooling water and compressed air systems, in order to enable equipment operation.
- For some fluid processes, 'wet commissioning' by operation with water, before introducing process materials. Often the pre-commissioning activity of plant flushing is combined with this for reasons of economy. Plants where the introduction of water is undesirable or hazardous may avoid this operation, incorporate a drying procedure, or utilize an inert fluid.
- Finally, introducing process materials to the plant, and building up to full commercial operation.

Commissioning requires a team of people with a background of plant design, plant operation, and plant maintenance. Quite often the consequence is that most of the construction engineers are not well suited to the task, and the plant design engineers are given this responsibility, which has a logical overlap. Some organizations employ specialized commissioning engineers, at least to lead the commissioning effort. This can be a very good investment, especially for a large plant where by providing for dedicated responsibility and focus, significant improvement on schedule and avoidance of negative start-up incidents may be achieved. A day extra taken up during commissioning is the same to the plant owner as a day extra taken up during design or construction; in fact it is likely to cost more, as the plant owner's commitments in terms of product marketing and operational costs are likely to be higher.

Because commissioning comes at the back end of the project there is a danger that the work may be under-resourced, because the funds have been pilfered to pay for budget overruns. It is essential to comprehend the scope and length of commissioning activities and include them in the initial project plan and budget allocations, and ensure that this commitment is maintained.

Detailed planning of commissioning and plant handover is as essential a component of the overall project plan and schedule as any other grouping of activities. Inevitably, some part of the critical path goes through commissioning up to handover. Like all planning, it is reliant on determining the methodology and procedures to be utilized, as well as the work breakdown, the activity durations, and the sequence logic which arise. These issues must be developed and agreed before it is possible to have confidence in the schedule leading up to plant operation.

When drawing up detailed commissioning plans and checklists, there are two important sources apart from previous experience:[1] firstly the process technology, and secondly the equipment vendors. With regard to the latter, it is necessary to review in detail the instructions provided in the operating and maintenance manuals, which must therefore be available well before the time of commissioning. The actual work of reviewing the manuals, clarifying any problems, and drawing up appropriate plans and checklists can consume many hours, but it must be done properly. Especially for high-cost plant items, it is often considered to be worthwhile to bring vendor commissioning engineers to site, thereby introducing an experienced specialist and reducing the load on the project commissioning team. Responsibility should be given to the vendor for ensuring that the equipment has been properly installed, thus

[1] Previous experience may be supplemented by industry literature such as the relevant API standards.

minimizing the possibilities that any future guarantee claims may be contested on these grounds. However, this can be a very expensive option. If the advantages are considered to warrant the expenditure, the requirement and associated service conditions should be included in the relevant equipment purchase enquiry, bid comparisons, and negotiations. If requested later, these services might be even more expensive, and the willingness to take responsibility as required tends to be remarkably diminished! Once again, advance planning is needed.

For the purposes of 'hot' commissioning, plants of all but minimal complexity have to be broken down into individual units or modules, which are commissioned in a sequence determined by the logic of operation and the need to build up intermediate material inventories. This logic in turn determines the availability required of process feedstocks, reagents, catalysts, utilities, and support services. As the whole commissioning sequence can occupy a number of months, it is obviously a matter of importance that the whole project schedule, working back into design and procurement activities, should follow the eventual commissioning priorities and that the corresponding plant groupings of modules and services should be understood.

Up to and including cold commissioning, and possibly wet commissioning, the activities are mainly those of discipline or equipment specialists, whereas the sequence and execution of hot commissioning is a process activity which is usually part of the process technology. As such, the work may be the responsibility of a technology provider, which is a separate entity from the rest of the project team. Any such split should not be allowed to result in separation of planning activities, unless the overall schedule duration is unimportant.

Since commissioning usually marks the handover of the plant to an operational organization, the formalities of acceptance and transfer (which may be quite onerous and time-consuming) have to be determined, and included in the plan. Other matters to be agreed at the project/plant-operation interface include:

- provision of operational and maintenance personnel and services during commissioning;
- provision of utilities, including electrical power and the associated interface organization;
- provision of reagents and lubricants; and
- determination of precise responsibility at any point in time for plant custody, operation, and maintenance (inevitably including various aspects of safety).

24.1 Safety during commissioning

Because commissioning is an engineering responsibility, and several specific precautions are necessary, it may be worthwhile to review a few issues which affect the safety of the operation. Competence is the first fundamental of safety. This implies that:

- work is properly planned, and
- everyone engaged in commissioning has clear responsibilities and has adequate competence and training to carry them out.

Following this generalization, there is a checklist of items which should be addressed in planning work and in work procedures.

- Communication must be adequate for the task to be performed and to cover any contingency that might go wrong. 'Communication' includes communication devices (radios, telephones, loudspeakers, etc.), signs ('Keep Out', 'Energized', 'Lubricated', 'Accepted for Operation', 'Do Not Start', etc.), written procedures, and vocabulary. It also includes frequent briefings as to what activities are planned, and their consequences, for those people involved, those affected (in particular construction workers), and those who might be exposed to hazards.
- There must be safe access to, and at least two exits from, any equipment item or plant area being commissioned.
- All safety devices and safety interlocks pertaining to a plant item should be made operational, calibrated, and proved before starting it. If this is not fully possible, say in the case of a turbine overspeed trip, then proving the device under controlled no-load conditions should be the first operation after starting. A witnessed checklist should be utilized for recording the proof of safety device and interlock operation.
- A switchgear lockout procedure, and the associated locks and keys, are required to ensure that an equipment item released for maintenance or modification is not started without the agreement of the person to whom it was released, and to confirm that the item is ready for operation.
- A procedure is required to formally record who has authority and responsibility for plant custody, operation, and maintenance at any time, as discussed above.

Chapter 25

Communication

It is evident that the quality of communication within a project is critical. The work is a team effort. One discipline's output is another discipline's input. The information transfer has to be co-ordinated in time and co-ordinated in content (to ensure that it is appropriate and comprehensive), and there has to be an effective feedback loop to ensure that information is understood and that perceived problems and conflicts are properly handled. Outputs of one discipline often affect the work, completed or still to be done, of other disciplines, in ways not identified or planned as prime input information. (For example, consequences of instrumentation design, see Section 18.5) At management level, a balance constantly has to be struck between the often-conflicting demands of time, cost, quality, and scope; for correct compromises, comprehensive information is essential. The larger the project or the more ambitious the time schedule, the greater the need for quick and appropriate communication.

Communication is thus an important project activity in its own right, and must be planned, resourced, and managed just as for any other activity. Communication is customarily planned by the following.

- *Circulation of documents.* All project documents should be included in a circulation matrix, in which each type of document (general design criteria, type of contract, type of drawing, specification, client report, etc.) appears on one axis, and each project discipline (lead engineer, expediting manager, chief field engineer, etc.) appears on the other. The nature of communication (if any) is detailed at the intersections. The 'nature' code may be: 'i' for 'copy for information', 'r' for 'copy to be circulated and returned with review comments', 'a' for 'for approval', 'c' for 'for construction', and so on. Communication matrix charts are a live procedure, updated in response to the needs of the team. They may be wasted unless there is a dedicated responsibility to ensure that documents go back and forth as ordained.

- *Control of external correspondence*, and specific procedures for specific types of communication. All external communication, particularly communication with the client, vendors, and contractors, needs to be the subject of procedures controlling formulation, authorization, record, and (where required), proof of transmission and prompt for follow-up.
- *Meetings*, generally scheduled weekly and monthly: a heavy but necessary drain on management time.
- *Reports*, concerning in particular: meetings; activities such as inspection, expediting and site visits; project controls such as cost reports; and project progress reports to the client.
- *Documentation indexes and project component registers.* The indexes enable users to know what documentation exists, and to access it. Registers are indexes to information which is not the subject of a specific document (such as equipment lists), but the nomenclature is not standardized: it is quite common to use 'register' and 'index' interchangeably. Equipment registers or lists, giving a summary of the main characteristics such as mass or power rating, are essential control documents referenced by all disciplines.
- *Briefs*, discussed below.

A brief is essentially an aid to planning an activity, and as such it is part of the basic project execution philosophy of 'Plan the work then work the plan'. Its objective is to ensure that all important aspects of a job are understood and communicated before the job is performed.

The effectiveness of briefs is enhanced by the use of checklists, which serve not only to remind the author of the brief's essential contents, but also to prompt him to ensure that he has himself completed all the necessary preceding activities. Here is a general form of checklist.

- Description of activities to which the brief refers.
- List of reference documentation (relevant information and data, work instructions and applicable standards, and acceptability criteria) needed for the task.
- Any special materials, equipment, skills, or methods to be employed when carrying out the activity.
- Any special hazards or safety precautions.
- Note of critical aspects of the activity.
- Note on interaction with other activities.

- Note on budget constraints, cost allowances, and schedule implications of the activity.
- Reports, notifications, and approvals required during the course of, or on completion of, the activity.

It has to be emphasized that a brief is a cover document rather than an instruction, its intention being to bring together information and advice which already exists in less focused form. Internal project instructions are made and issued in the form of project strategies, design criteria, procedures, specifications, requisitions, and so on; external instructions are made in the form of purchase orders, contracts, and formalized work orders and amendments issued in terms of the orders and contracts. It is not wise to have any parallel system by which instructions may be given. At the least, this can disrupt management and cost control, but it can also generate serious confusion.

For groups of tasks, such as the commencement of a site construction contract, the brief is in effect the agenda for the 'kick-off', and should have as many parts as there are contributing authors to the contract, usually at least two (technical and commercial).

By the nature of a brief, it is a very useful document when checking the performance of an activity, and thus a formalized brief document is a useful quality assurance tool. The following is a list of activities for which formalized briefs should be considered, including some specific checklist items for the activity.

- Drawings:
  - existing standard designs and previous project designs, on which the design may be based;
  - relevant specifications and engineering calculations and sketches;
  - equipment vendor data and drawings;
  - impacting activities of other disciplines.
- Inspection activities:
  - critical features, dimensions, and other acceptance criteria;
  - use of templates for testing match of separately manufactured parts;
  - concerns relevant to the particular manufacturer;
  - assembly or disassembly and protection for shipment.
- Field erection of item:
  - state of assembly or disassembly for shipment;
  - size and mass of major assemblies;
  - special lifting instructions;
  - site storage instructions;

- special erection equipment and tools;
- 'consumable' items such as packers, bolts, special lubricants, and cable glands to be supplied by others;
- vendor installation manual;
- site attendance of vendor specialist;
- special installation instructions;
- sequence-of-erection constraints.

- Commissioning of item:
  - checklists for acceptance of erection and for pre-commissioning functional tests;
  - instructions for cleaning and flushing;
  - vendor manual, including operating and commissioning instructions;
  - lubrication schedule;
  - spare parts ordered for commissioning and operation;
  - safety considerations and hazards;
  - performance tests and measurements required;
  - vendor telephone number.

Chapter 26

Change and Chaos

26.1 Change

We started our review of project management, in Chapter 3, by stressing the fundamental need of planning and working to plan. Inevitably there will be a need to change the plan. It is unlikely that every event will materialize as foreseen, or that arising from the unforeseen, there will be no previous decisions or designs that subsequently are seen to be wrong or sub-optimal.

As we have seen, there are indeed bound to be many assumptions made during planning that will inevitably lead to some measure of revision and re-work. The issue is not therefore whether to permit any changes, but how to control them so that they do not destroy the manageability of the project.

Change control has a few components: change recognition, change evaluation, change approval or rejection, and change implementation, including the consequent revision to plan. These are discussed below.

- *Change recognition* may sound an odd subject to the uninitiated, but it is a fact that changes are too often made inadvertently or surreptitiously, with damaging consequences. An error is in general a form of inadvertent change from what was intended. Once an error has been made, it may be discovered and consequently the relevant work may be surreptitiously changed, with unforeseen repercussions.

  Process plant design and construction is such a heavily interlinked exercise of different activities and disciplines, that the consequences of change are very easy to underestimate. In fact 'change' is invariably one of the most emotive issues in the business.

  Change recognition is mainly dependent on the training and conscientiousness of the work performers and their supervisors. It is essential to have a climate in which it is unacceptable to carry out changes without formal notification to the other team members, or without approval subject to agreed procedures. Of course, a draughtsman or engineer is still entitled to use his eraser (or delete-key), and that does not automatically constitute a change as envisaged. It is only a project change if the document has already been communicated to someone else, and this distinction must be emphasized. The prevention of issue of drawings, or indeed any documents incorporating revisions, without any change to the revision status is one of the perennial management 'change-recognition' concerns. Appropriate programming of computerized document control systems can be very effective in overcoming the problem.

- *Change evaluation* is the process of communicating the proposed change to all disciplines that may be affected, getting their feedback, and evaluating the full technical, schedule, and cost impact of the change. There is some skill and experience involved in ensuring that all affected parties and work are included. This process may be greatly improved by the meticulous record and recall of information use, but this ideal seems to be quite difficult to maintain in practice – much information is passed on informally.

  Following the receipt of feedback comes a process which may demand as much engineering knowledge and ingenuity as the original plant design: the development of ways of implementing the change and overcoming negative consequences. Once these are understood, there is a process of evaluation and optimization leading to a decision on the best way of handling the change, and an estimate of the associated costs and schedule implications.

  The implications of a change to all facets of the project work are usually called the 'impact'. It has long been realized that the impact is much heavier in the middle of the design process than at the beginning or end. At the beginning, when designs are conceptual rather than detailed, not much other work is likely to be affected, and users of information understand that it is provisional and subject to confirmation. At the end, the implications of change are much clearer and work currently in progress is not affected. As a result, when estimating the consequences of change, several organizations apply an 'impact factor' (ranging from 1 to as much as 4) to the first estimate of additional manhours, depending on the stage of the project.
- *Change approval or rejection* is a management or client decision based on the change evaluation and consequent recommendation.

At this stage there may be disputes about whether a change is a 'change' for which the client bears responsibility, or an 'error correction' for which the project team bears responsibility, and a dispute resolution procedure may be required. The project engineer's attention is drawn to the strategic precepts outlined in Chapter 7 on how best to handle such situations, the most important of which is to be well-positioned before the conflict arises by meticulous project scope control and the elimination of 'weasel words' in design criteria.

- *Change implementation* has to be rigidly controlled and followed up. The follow-up system has to ensure not only that all relevant project work is revised accordingly but also that all consequent document revisions are received and utilized by all who had the previous revision, which must be destroyed or visibly cancelled.

Change on a project of significant size cannot possibly be managed without adequate procedures that embrace change recognition, change proposal notification, change evaluation, change approval, change implementation, and document revision. As part of these procedures it is inevitably necessary to maintain numbered registers of change proposals, change notices (which record the evaluation and recommendations), and change orders.

The work of evaluating a proposed change can be extensive and can impact heavily on scheduled work performance. And yet anything to do with a change has to receive priority: delays tend to greatly increase the impact of implementation – more work has been done that has to be changed. It is therefore customary to follow a two- or even three-part procedure for change control (whereby quick decisions can be made as to whether to proceed with more formalized evaluation or whether to drop the proposal) and to allow certain categories of minor change to be processed informally, subject to appropriate authority level.

26.2 Chaos

The following message may be found displayed at the workstations of many process plant project engineers:

> 'From out of the chaos, a small voice spoke to me and said *Relax and be happy, things could be worse.* So I relaxed and was happy ... and things got worse.'

The fight against chaos is a subject of constant awareness by even the most methodical of process plant engineers, however good the systems employed and however well-drilled the project team. Chaos takes many forms, but the general result is an unpredictable working environment: documents not available on time, suppliers and contractors kept waiting for replies, wrong and contradictory information circulated, and so on. The cause is clearly that the project team has become overloaded in its capacity to deliver the right goods at the right time. The effect inevitably includes not only reduction of quality, which in the case of engineering essentially means too many mistakes or omissions in the information issued, but also failure to meet schedule commitments. There also inescapably follows an overrun in both direct field costs and the duration and cost of engineering itself, as efficiency levels deteriorate. In fact as the effects build up on each other, it is easily possible for the project to spiral out of control unless firm management action is taken.

Fig. 26.1 Chaos

Project teams can be chaotic from the start, almost irrespective of the load placed on them, through failure to properly plan and organize their work, for example because the complexity of the project is underestimated, because concepts are not frozen before commencing detail work, or because of inadequate competence. The ability of a project team to deal with complex tasks *when stretched to their limit* is a quantitative matter. On a scale 0–100, for different teams there will be different scores, depending on the teams' training, individual skills, systems employed, and quality of management.

The load placed on a project engineering team is the demand made on the team to produce quality information on time. As we have observed in Chapter 7, The Contracting Environment, various circumstances surrounding the creation of a project, and in particular a competitive environment, may combine to set the project team a challenge – a load – which even at the outset may be at or beyond the limits of its capability. On the other hand, where little competition exists, quite a pedestrian performance may be sufficient. Some elements of the challenge may be comparatively quantifiable, such as the manhour allowance per drawing or purchase order of a particular type. Some are not so simple, for instance the influence of parallel activities, assumptions and consequential re-work (which are necessitated by the schedule compression) or, generally, the number of design iterations needed for whatever reason.

On top of the load placed on the project team at the project set-up stage comes an additional load placed by changes and events beyond the control of the team. These include the following.

- Process changes, changes to the process flowsheets, or changes in fundamental process information, typically made because of technology development, the late appreciation of better technological possibilities, changed site conditions (feedstock, operational needs, costs, etc.), or changes to the product market.
- Other design criteria changes – usually driven by cost or time 'surprises' – when unexpected consequences of specification and design choices become apparent at the procurement stage, and changes are required in order to maintain the project budget or schedule, or to take advantage of opportunities for improvement.
- Changes required to counteract the effects of poor performance by suppliers or contractors; for instance if an equipment supplier has to be replaced by another, the adoption of proprietary equipment designs different from those used for the design of interfacing structures, electrical supply, and connecting piping.

- Correction of errors made in the conceptual work undertaken before project commencement, or in the project work as it unfolds.

In summary there are two components to the load placed on the project team: the load of producing initially defined information (document production, follow-up, inspection activities, and associated reporting), and the load of changing or adding to the information and its basis as the need arises. Some components of the latter may be anticipated and allowed at the project planning stage, others may not (or can not) be anticipated.

Returning to the subject of overload prediction, and defining the project team 'workrate' as the load of work per unit of time, we have from the above the basic equation

$$\text{Workrate required} = \frac{\text{initial WBS}^{1} + \text{changes}}{\text{schedule duration}}$$

This must be compared with

$$\text{Workrate capacity} = \text{team competence and motivation} + \text{resource increases}$$

If the project team, by whatever reasons of competence, experience, and good organization, has a workrate capacity which exceeds the planned requirement, in a challenging environment there are various factors which tend to cause the workrate required and the capacity to move into equilibrium, outlined below.

- As the work advances, the client may demand more changes, seeing the apparent ability of the team to cope with the inevitable possibilities of design improvement. These changes may well be initiated by the project team members themselves.
- There may be a desire to use up more time in commercial negotiation in order to get better deals from suppliers and contractors, thereby effectively compressing the engineering schedule to maintain the overall project schedule.
- Project resources may be diverted to other work that seems at the time to be more pressing.
- The project team motivation may be reduced, typically by a corporate executive outside the project team who sees some opportunities to

[1] Work breakdown structure, as planned.

make petty economies by cutting benefits, or by not maintaining them in an inflationary situation.

However, there are environments which are not challenging, where changes are hardly allowed, resources are not diverted, and plans and commercial interfaces are generally quite rigid, most normally in government projects. But even here, capacity may be reduced as project team members (especially the more capable ones) depart in search of a more challenging environment.

However high the initial capacity, the usual situation is that the workrate capacity and workrate requirement rapidly come to an equilibrium, which requires management effort for its further maintenance. Chaos can be expected if the workrate required exceeds the capacity, although most teams will rise to the occasion if this condition is not too prolonged. Nevertheless, certainly there is an overload intensity and duration at which performance begins to fall off exponentially.

Because of all the variables involved, it is not easy from a project management viewpoint to predict the precise workrate at which chaos will escalate. Rather it is necessary to look for the signs of impending chaos, and then know that urgent management action is necessary, in the form of the following.

- Accept less changes.
- Decrease the initial WBS; shed off some of the load, for instance peripheral parts of the plant, to another team.
- Extend the schedule, that is, the real engineering schedule – not the apparent schedule, which does not account for information delays arising from extended commercial negotiation, late client decisions, poor vendor performance, and so on).
- Add more resources or improve their quality – not easily done in the middle of a project, owing to the learning curve. If resources have been diverted, it would be far better to recover them.
- Improve motivation. (Fire that bean-counting executive!)

The 'signs of impending chaos' include the following.

- An increasing trend of targets missed on weekly progress review against *current* targets (that is, after allowing for any rescheduling).
- A high percentage of effort expended on changes and re-work, at a time when output of planned work is behind schedule.
- An increasing trend of delays in responding to technical or commercial queries (for example, take the site query register, look up the

outstanding unanswered queries and the days outstanding for each, and add them up).

- Less 'buzz' in the office, more cynical remarks, more graffiti, and so on. People do not like working in an environment where they seem to be losing.

It must be evident that a state of chaos is a no-win situation for the engineer, and for the project as a whole. There is no silver lining to this cloud, no upside benefits to trade against the downside. Efficiency goes down, quality goes down, and project costs and durations go up. And yet some clients, or rather their representatives or other management of the various parties involved in a project, sometimes almost seem to be conspiring to wish these circumstances on the engineering team in an apparent attempt at group suicide. Possibly this comes about as a result of ignorance, or maybe from a misguided attempt to maintain a position of dominance over the engineering team. An engineering team that is past the threshold of chaos has no scope for initiative. It operates in a purely reactive mode, is disinclined to strategize its relationships, and is easily manipulated.

Some engineers find it very difficult to escape from this situation, most frequently when the client or dominant party is both experienced and cunning, and typically lets out the rope of workrate adjustment in carefully measured little increments, enough to stave off disaster but never quite enough to let the engineers get right on top of their work and regain the initiative. Engineers who are more successful anticipate the day when schedule extension or other action may be necessary, and state categorically what is needed in order to continue to accept the responsibility for plant safety, and the engineering quality on which it depends.

Chapter 27

Fast-Track Projects

Initially, a project activity sequence network may be prepared by following the logical sequence of individual activities being performed when all the required information, or previous related activities, are completed.

1. Equipment is ordered and vendor drawings are obtained.
2. Based on these drawings, the steelwork supporting the equipment is designed.
3. Based on the steelwork drawings, foundation plinths are designed.
4. Pile designs are prepared.
5. The piles, pile caps, and plinths are built, steel erected, and equipment erected.

However, a project schedule based exclusively on such an orderly sequence of events is seldom acceptable. Having identified the critical path (or, after sufficient experience, usually knowing instinctively what has to be done), it is necessary to reduce the overall planned duration by various means.

In Section 4.4, Co-ordinating engineering work, we discussed the most basic methods of shortening the schedule, including in particular the adoption of standard (or existing) designs. These may involve some changes to the process flowsheet, and the elimination of uncertainties by setting conservative design or equipment selection parameters. There remain two very common devices to shorten the schedule of engineering work: parallel working and assumption of design information. In their application the two devices are almost identical.

In parallel working, two or more interrelated design activities are carried out simultaneously. For instance, the structural steelwork and piping are designed simultaneously based on the layout drawings, rather than the piping being designed to suit the steelwork after its structural design, as in Fig. 4.1. The efficiency of this process is dependent on such factors as the quality of teamwork, the design aids available to assist co-ordination, and the success with which the layout drawings have anticipated the final structural design.

The practice of parallel working extends into the manufacturing and construction phases. Examples are:

- equipment design and manufacture commencing, while secondary equipment design information, such as positions of piping connections, is still 'on hold';
- placing of contracts for construction work before the drawings to define the details of the work have been completed; and
- the construction of earthworks proceeding before the plant layout details have been finalized.

In the making of assumptions of design information, usually related to information from a third party, the progress of design work is based on the best (or sometimes most conservative) estimate of the information. For example, structural steelwork design may be based on the anticipated mass and dimensions of a supported equipment item for which final vendor's information is not yet available. Evidently the accuracy of the assumptions, the designed-in flexibility to accommodate incorrect assumptions, and the ability to influence the equipment item vendor to comply with the assumptions (or the freedom to choose a compliant vendor) will affect the efficiency of this process.

If the assumptions to be made relate to the performance of work by another project engineering discipline, say if the mechanical engineer orders a pump before the process engineer finalizes the calculations which will affect the required performance, it is clearly only a matter of semantics whether the practice is described as an assumption or parallel work.

If work proceeds in parallel or by assumption, there is clearly a possibility (for a large amount of such work, a certainty) that there will be some errors and inconsistency. Time and resources must be allowed to detect the problems and correct the work in the design office, redesign the equipment being manufactured, modify the construction on site, or a mixture of all three.

The practices outlined above are effectively a means of breaking the logic of schedule dependency; there are also other ways of going about it. One method quite often practised at the civil/structural-steel interface is to work 'from both ends towards the middle'. For instance, there may be a piece of process equipment to be supported; information about the precise interface dimensions and holding-down bolt arrangement is dependent on getting final vendor information, which is itself dependent on a whole series of events, including order placement. On the other hand, there is a critical requirement to commence work on the foundations, which as outlined above are designed to suit the structural steelwork, in turn designed to suit the supported equipment. The dependency may be broken by deciding instead (*if it looks to be a reasonable technical solution*) to design the structural steelwork to suit both the foundation and the equipment. This makes it possible to get on with the foundation design and construction without waiting for the equipment vendor's information.

The last example will be seen to be one of very many possible variations: for example, designing the steelwork and/or the foundation in such a way that it can accommodate a sufficient range of equipment dimensions, or arranging for the foundation/steelwork to be adapted on site as may be required.

In fact, on closer examination, all these methods of shortening the schedule are seen to be one and the same action, which is to reject the limitations of sequential activities, stipulate that certain activities will commence at the time required to suit the schedule, and then devise a plan to overcome the consequences. Clearly such an approach can be taken too far, when everything is ordered and sent to site without much design integration at all, to be 'fixed up on site'. This is not recommended; but from a philosophical point of view, in search of the best plan, it may be just as instructive to start from this 'ultimate disorder' scenario as to start from the 'logical sequence' network.

Summarizing the above, the sequencing and scheduling of engineering work is rarely a simply determined process. It usually includes compromises to accelerate the work by reconsidering design and procurement parameters, by parallel working, by making assumptions, and by generally challenging the initially adopted sequence logic against the consequences of acceleration, as reflected in increased design error and additional site work.

The difficulty of deciding on the best compromises with regard to parallel working and assumptions increases exponentially with the size and complexity of the plant being built, principally because of the knock-on effect of design revisions and errors. The engineering manager is well advised to be more conservative (in estimating the consequences of engineering schedule compression) when planning larger projects.

To illustrate, in a project and industry background, some of the problems and opportunities which can arise, here are the outlines of two projects where a fast-track approach was adopted.

27.1 Case study A: Fabulous Fertilizers' new plant

The client, FF, predicted that the market for its product would become greatly improved in the next northern-hemisphere season, but to take advantage of this opportunity, a major capacity expansion had to be commissioned within 12 months. The last similar plant construction had taken 16 months to complete. FF's project engineer studied the schedules and close-out reports for similar projects, and concluded that he could knock 4 months off the last project schedule if the following conditions were met.

- *Procurement of the major process equipment items began immediately*, based on extrapolation of previous project data (without waiting for project-specific flowsheets and mass balance to be finalized). The basic data for the equipment would then become requirements for flowsheet and data sheet preparation, rather than vice versa. Design details arising out of plant layout could still be accepted several weeks after purchase commitment.
- *The longest delivery item*, the reactor, was ordered from the supplier for a previous project at the same price per ton, thereby saving the time for competitive bidding.
- *The plant was separated into three distinct modules*, instead of the previous practice of a single integrated unit. This would enable him to concentrate his (limited) design team on the design of the reactor section without regard to the other two simpler modules, whose delay could be accommodated. The penalty for this was two additional piperacks, extra pipework, and a need for more ground, but there were also other beneficial spin-offs: separate and less expensive construction contractors could be used for the two smaller modules, and the feed preparation module could be commissioned and used to build up intermediate feedstock, while it was still safe to carry out hot work on the reactor section.

No other short-cuts were required to FF's normal project practice, and the project engineer decided to maintain all his normal project procedures rather than risk the consequences of impromptu short-cuts. The project engineer took the trouble to discuss the whole philosophy with FF's commercial manager and procurement department. They agreed with him on an absolute maximum 10-day period for the assessment of equipment bids, and 1 week further for negotiation and order placement, and also suggested that a few traditional bidders be eliminated because of a record of tardy performance.

FF got their plant commissioned within the 12-month period – but not without additional, unforeseen costs and problems. Construction plans had to be changed a few times, as insufficiently thought-out designs had to be modified in the field, but it was no coincidence that the errors occurred in the less critical areas where there was some flexibility to make changes. And undoubtedly, some premium was paid for abbreviating the procurement deliberations and for excluding vendors and contractors who were adjudged unable to meet the schedule (although most of them claimed otherwise). The extra costs of fast-tracking the project were considered to be far less than the commercial benefits of early product delivery.

27.2 Case study B: Magnificent Metals' new gold plant

Magnificent Metals, MM, is a very big organization that owns many gold mines, mainly designed by their own projects department, headed by Ivan the Terrible. The projects department had become a very expensive institution, which took a long time to deliver plants that embodied many unnecessarily expensive features. Ivan's new managing director, Smart Alec, decided that MM's new venture, Soggy Swamp, would be designed and built by contracting organizations and not by the projects department. Ivan's role was restricted to that of owner's engineer, responsible for specifying the plant and supervising the contractor. The shareholders were informed of the venture and its commissioning date, set to coincide with the expected upturn in the gold market in 3 years' time.

Ivan commenced the specification of the project works, but was beset with difficulties, principally that the ore mineralogy and grade were found to be somewhat uncertain. Ivan decided on a campaign of ore sampling and process testwork. He had made his reputation by getting things right – not by taking chances (he has no understanding of the word probability) – and he took over a year to finalize the basic process design parameters. Alec became increasingly impatient, and perceived that he might be embarrassed by failing to meet the committed completion date. But with only 18 months in hand, the plant was still insufficiently specified to permit a lump-sum bidding process by the contractors – Ivan was quite adamant that only rubbish would be delivered if this was attempted. He needed to be able to participate in detail design decisions. It was therefore decided to engage in a competitive bidding process to appoint a managing contractor, who was to be payed a fixed price for EPCM (Engineering, Procurement, and Construction Management) services. All direct field costs would be reimbursed by MM.

The winning bidder was the Always Adventurous construction company (AA). By this time there were only 16 months left until completion of commissioning. The project had become fast track, something of which Ivan had no experience. 'No problem', said AA. 'We have done this before – we commissioned the Desert Gold plant in under 16 months, and look how well it functions'.

Now we will not belabour the reader with all that went wrong with the Soggy Swamp project. It was a disaster: never on schedule and, due to the ensuing pressures and ill-considered short-cuts, also technically unsatisfactory and with a cost well over budget. We will rather dwell on the differences between Desert Gold and Soggy Swamp, as Alec and Ivan finally recognized them in hindsight.

The main difference was the degree to which design basics had been finalized before the contract award. AA were awarded the contract for Desert Gold after a detailed study characterized by the following.

- Flowsheets had been developed to a stage where the mass balance could be frozen for all the main process streams. Not so for Soggy – 2 months into the project schedule, AA's process engineers were still arguing the interpretation of the process testwork (not all of which had been divulged in the bidding process) and the consequences to the flowsheets.
- All the main equipment design parameters had been set, bids had been solicited accordingly, and vendors chosen. At the outset of Desert Gold, purchase orders for the main equipment items could immediately be negotiated with the vendors, based on the previous bids. Not so for Soggy: bids had to be solicited from scratch, and first the process data had to be finalized.
- The Desert Gold layouts had been extensively reviewed and optimized, and coarse-scale Hazops had been held for the more important plant areas. The Soggy layouts were being reconsidered well into detailed design, and when AA were finally compelled to carry out Hazops ('Hazops for a gold plant? We've built ten of them without any problems'), Ivan's team felt obliged to require some major re-work.

Apart from physical differences such as starting from a different base, there were equally important differences in personnel and human relations. The lead engineers on Desert Gold had all been involved with the study and got to know the owner's supervisory team, which consisted of only four individuals. Personality clashes and differences of technical approach were behind them. Moreover, the Desert Gold owner's project director had insisted that their project manager and his supervisory team be given significant incentives for meeting all the project objectives. No such luck for Ivan's 20-man team. They came from the projects department's rigid background and ideology, and insisted that they must check every design detail and obtain perfection. They received no reward for completion on time, and saw only the probability of criticism if the design was imperfect. It may be remarked with surprise, then, that some of the design work turned out to be very poor. The answer is that one of the most important factors is co-ordination, and a plethora of supervisory specialists does little to help that, and often much to hinder. Another important factor was that it became impossible for AA to keep all their best engineers and technicians working on the Soggy project – many resigned rather than work with the 'Terrible Team'.

By Ivan's requirement, the AA contract included rather punitive damages for the consequences of design error. The AA team were totally disinclined to make any assumptions that might accelerate the project, but rather dissipated their ingenuity in writing clever letters to their client, pointing out how the client team was alone responsible for all the delays and re-work.

Now, we have labelled Soggy Swamp as a disaster, but that is only because we have information that is otherwise privy to Smart Alec, Ivan, and the project teams. In fact Alec's reputation could afford no such disaster, and when (a few weeks later than the planned commissioning date) he understood how bad the outcome was, he initiated a public relations campaign drawing attention to the completely unforeseen difficulties arising from the unique features of ore mineralogy. He diverted funds from other allocations into the project coffers, and embarked on an expensive plant upgrade (for which AA were paid handsomely). Eight months later, he trumpeted the success of Magnificent Metals' experts in overcoming all the problems, and AA had another good project reference to their name, enabling them to win the bid for Happy Half-wits' new process unit.

Chapter 28

Advanced Information Management Systems

Project engineering and management consists of a number of types of individual activity, surrounded by a mass of information flow, which connects the activities according to certain procedures. The scope for enhancing project performance by utilizing advanced management information systems has been obvious since more powerful computers became generally available. Many such systems have been and continue to be developed. A few of the typical features and aspirations follow.

- All information flow may be set up within the computer network by paths that reflect the project procedures and workflow.
- 'Intelligent' P&IDs may be set up, from which information such as equipment lists and pipeline and instrument data are automatically transferred to those disciplines. All information regarding a P&ID item may be accessed by clicking on the item.
- Equipment data sheets, purchase orders, and contract schedules may be formatted on standard templates which reflect the ordered, single-entry-point information flow from one discipline to another, with the required approvals built in to the route of passage.
- The information systems can be combined with advanced plant modelling systems, and interface to external software packages, such as for process modelling and pipe stress analysis.
- Before a drawing or model is commenced, the associated input of specifications, calculations, preceding and other-discipline drawings, and equipment vendor information may be made available electronically to the designer without need to identify the inputs and collect them each time they are needed.
- The more mechanistic drawing processes, such as electrical and

instrumentation lower-level diagrams and schedules, may be generated from the P&ID and engineering database without much designer input.

- Drawings may be automatically compared with design data such as P&IDs.
- Documents may be electronically circulated for review by a defined circulation list according to document type, and may not be issued or revised without the appropriate approvals.
- Project control schedules may be automatically collated, for instance by electronically linking cost schedules to information in the purchase order.
- Reports of all types may be formatted as the user may require. Cost reports may include breakdown by area, discipline, activity, or whatever a specialized user requires. Likewise, progress reports, discipline productivity reports, quality check records, and cash flow reports and forecasts can be generated from the same database.
- Engineering managers may receive at the beginning of each week a status report on work required according to project plan, critical outstanding items, and changes made and under consideration – all quickly and automatically.
- Design databank, standard calculation software, and back-up information may be made electronically available in the context in which they are utilized.
- A list of affected documents may be produced automatically when a change is made to any project data.
- Materials control may be initiated from the single-point entry of an item on a drawing or schedule, and from there automatically taken through the stages of roll-up, enquiry, purchase (and, maybe separately, addition to construction contract schedule), inspection, despatch, payment, site stores control, and information recall for maintenance purposes.
- No paper need be utilized for any of the above, and not only the documents themselves but all approvals, issues, and revisions will be securely recorded, and will be made available in a highly organized and accessible form for the life of the plant.

The most fundamental aspect of integration is the datacentric system, which creates a single-entry database as the heart of the project's information. Documents are no longer data archives; they are simply a focused extract from the database at a point in time. A general example of the major system components and their interaction is shown in Fig. 28.1. Software providers who have more or less developed such

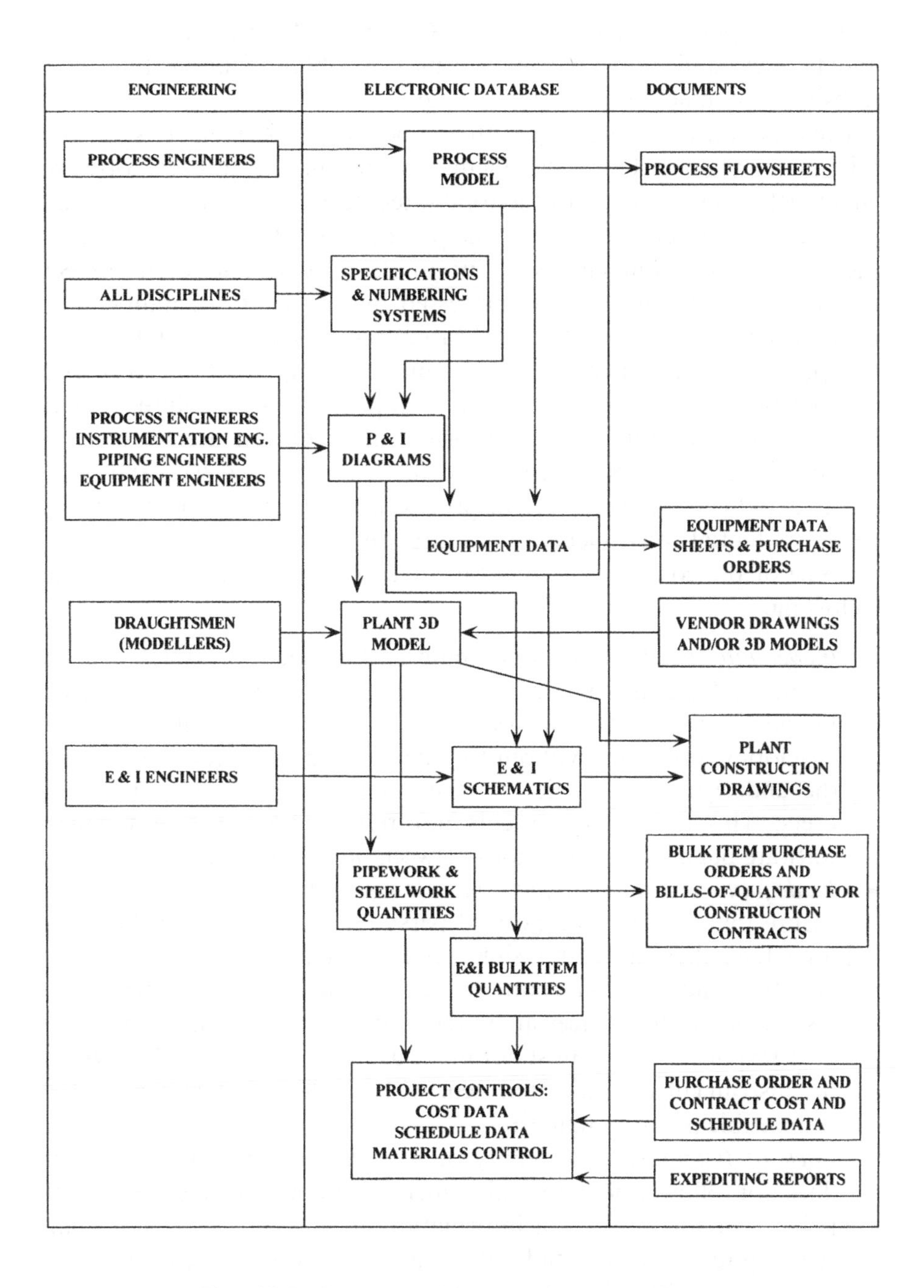

Fig. 28.1 Datacentric information management

systems (or parts of systems) include Cadcentre, Intergraph, and Rebis and Bentley, and more information may be obtained by visiting their websites or contacting their salespeople.

Most organizations dealing with engineering projects have, since the take-off of modern IT development, invested in and utilized software which delivers some parts of the total integration, and have experienced pressure for further integration. The potential benefits are obvious, and the expectation from clients is inevitable. Clients and even project managers may be profoundly impressed by the wealth of multicolour charts generated. And yet the application remains limited in relation to the potential, and there are many, many examples of projects and organizations that have experienced 'computer-assisted failure'; this is likely to continue for as long as insufficiently critical users are to be found. These computer systems need to be approached with the same attitude and caution as a piece of process equipment, except even more so: the potential consequences of failure are vast.

Some of the common reasons for unacceptable performance are the following.

- *Inadequate specification.* A computerized information system can be no better than the logic and demands which it is set up to reflect. Despite any number of warnings, organizations will persist in developing programs without first developing and rigorously testing the underlying flow logic and the adequacy of the outputs to all users. This cautionary note is not restricted to custom-designed software but also to connections between proven proprietary packages.
- *Inadequate control of inputs and changes.* It is absolutely vital to ensure that only the authorized personnel have access to the work for which they are responsible; that no unauthorized changes are possible; and that all changes are recorded. For example, if an instrumentation engineer alters the data of an instrument for which he is responsible, this must not make a change to the P&ID to which the instrument data is connected, without the authorization of the responsible process engineer.
- *Lack of flexibility.* The reality of a project organization is that it does not follow strictly logical sequences of operations, but exercises judgement when pieces of information are missing or are not final. Assumptions may be made to advance the work, based on taking a view that there is a reasonable probability of being right and that there will not be too many adverse consequences from being wrong, etc. It is *very* difficult to design a system that allows such

intermediate choices to permit work to go on, governed by a suitable level of extra surveillance and partial revision, which for instance a draughtsman may carry in his head as he develops a design.

The classic example of the project organization hamstrung by such lack of flexibility is where every purchase order is delayed interminably, then issued in the form of a letter of intent (bypassing the computer system), because the input information is never quite right for the computer. Accounts are then unpaid because there is no purchase order, the suppliers cease to perform, and so on. The project team blame the computer system, the system designers blame the project team's inputs, and frustration continues.

The flexibility issue can also be severe in relation to the control and numbering system demands of different clients, and the financial auditing requirements in different countries.

- *Underestimation of data input demands.* (The old saying: 'Rubbish in, rubbish out'.) The more complex and far-reaching the computer program, the greater the demands for complexity and accuracy of data input. This requires training, time, and checking. The checking routines have to be rigorously developed and maintained, with systematic analysis of the probability and consequence of error. Otherwise the system rapidly crashes and everybody blames the lowest level of employee, never the system designer.
- *Unforeseen consequences of input error.* Developing the previous theme, it is frequently experienced that certain types of input error have unforeseen results. For instance, a letter 'o' instead of a zero may result in a whole item of input disappearing from the system, which may only be realized when an entire pipeline is found to be non-existent during the course of pre-commissioning checks. Ultimately, such unforeseen problems are only eliminated by experience of the application.
- *Lack of training.* Projects, being discontinuous phenomena, are naturally associated with discontinuous employment and the use of new staff. Training needs and the associated time-lags easily lead to the use of relatively untrained staff over peak periods. All project personnel are involved, not just data input clerks. For instance, a site storeman who understands neither the need for meticulous input nor the consequences of error – but does understand the need to issue materials promptly under pressure – can and will reduce a computerized materials management system to irreversible chaos.
- *Unreliable software support staff.* Software developers may understand that they are well-positioned to make themselves

indispensable, and take advantage of it. Frequently they are able to talk their clients into financing customized developments, which are then owned by the software developer but on which the client that financed the development is reliant. The consequences may include extortionate rates for support services, and an arrogant attitude to the software users, who are usually well down the organizational tree from the executive who set up the deal. As a further consequence, the users may adopt an unconstructive attitude to 'making the system work', and on we go to chaos and disaster.

  There is another form of software disaster, in which the key software support staff simply become ill, meet with an accident, or disappear, and are found to be irreplaceable, at least within the needs of the project schedule.

- *Underestimation of cost and obsolescence.* Looking back on history, it appears that rather few complex computer applications have paid their way in commercial enterprises until the particular development came into widespread use. The normal development scenario is that of a gross underestimation of development and application costs, and an overestimation of effectiveness by everybody ... especially the salesman. Projects, as opposed to continuous-product organizations, are at a natural disadvantage here. Each project has its different needs, and there is no time for system development on the job.
- *Insufficiently 'robust' software*, which 'falls down' if an input error is made, an illegal operation is performed, or simply because of an inherent bug.
- *Inability to monitor or forecast progress adequately*, because of changed work methods. In particular it is no longer easy to measure progress by documents; it is instead necessary to consider objects. In addition, it is, typically, easy to underestimate the 'front end' of design, when the comprehensive system of specifications is set up.

In conclusion, the scope of what can be done for the project and engineering organization by advanced computer systems is truly wonderful, but the prudent user needs to treat all such systems critically. Utilize only those which are rigorously specified, tested on real applications (by real project people), not over-ambitious, and manifestly will pay their way in terms of a conservative business plan. There are definite cut-off points in project size and complexity below which it is not economical to operate fully integrated systems. Systems application below this threshold should not be attempted unless there are other non-economic factors such as a client who requires them (usually, for interface into overall plant information management systems).

Final Cycle
Strategies for Success

Chapter 29

Project Strategy Development

It may seem odd that this subject, which is the essence of deciding how to go about the whole project, should be left to the end. (The final chapter which follows is essentially a summary.) The reason is that it cannot be addressed economically until all the main issues have been exposed and discussed.

The project strategy is the name given to the highest level of a project plan. This word tends to be rather misused or abused in management literature. What it has in common with the military usage, from which it is derived, is that it addresses an entire plan of campaign (the project). Any specific issues, such as construction or engineering, should only be addressed in so far as they directly relate to the overall objectives and cannot be settled in a smaller forum. The introduction of too much detail, or too low a level of planning, inevitably detracts from focus on the major issues.

For instance, consider the means of addressing design criteria (the highest level of engineering planning) during project strategy development. With the exception of the most fundamental issue, the overall plant performance definition, the design criteria should be addressed as to how they will be formulated (which will be discussed below) rather than their content. There may well be lower-level strategies than the project strategy. For instance, it may well pay to organize a multi-functional strategic meeting to discuss the way of formulating and negotiating a particular construction contract within the overall project. The usual objectives are to ensure that the important engineering, procurement, and construction implications are understood and to derive an appropriate overall strategy, before starting any one function or discipline's work. However, such supplementary strategizing is conducted with the overall project strategy as a base; it is probably a misuse of the word strategic, but commonly employed when one wants people to 'think big'. 'Big' is of course always relative.

In the nature of organizational structure, there will be a strategy (stated or implied) for the performance of each independent party, and that party is unlikely to perform better than is allowed by the constraints of his planning ability. If a project is set up as a united owner-paid team, with a reasonable structure of partnership by which all team members participate in the fortunes of the project, one can expect a single strategy. In the case that the project is set up as a lump-sum turnkey operation to be undertaken by a single overall contractor, there must be at least two strategies: the owner needs his strategy for setting up and awarding the contract, managing the contractor, and getting the best out of him; the contractor needs his own strategy for project execution, including the management of his relationship with the owner, and for maximizing his returns within the scope offered by that relationship.

The fundamental need for strategic thinking for projects lies in the basic *modus operandi* for effective project work, set out in Chapter 3: plan the work then work the plan. Work is carried out effectively if it follows a plan that is subject to minimal revision and minimal questioning as work proceeds. All issues which might otherwise lead to a change of plan need to be anticipated and dealt with 'up front'.

These are the essentials of strategic planning, therefore.

- Start from as broad a base as possible and deal with issues as widely as possible. Any issues which are important to the project, including opportunities outside of the defined initial objectives, need to be addressed.
- Make the plan clear. Clarity does not allow the 'fudging' of any issues which could not be decided. It does not include unqualified words like 'best', as in 'best practice', which simply puts off the decision in deciding what is best. So for instance, instead of the unacceptable 'best engineering practice' (weasel-words for a manipulative type of specification), use rather 'engineering practice as determined by limiting return value analysis', or better still, 'engineering practice as per project manual No. XXX', provided of course that such a narrow definition is correct at the strategic level. There may be a need to identify some aspects of manual No. XXX that have to be subjected to a process of challenge and modification for the project, in which case the definition of that process (but probably not the outcome) can go into the strategic plan.
- Make the plan inclusive. There will be many decisions that cannot be taken at the start of the project due to lack of information, but

there should be set into motion a process by which the necessary information will be defined, obtained, and a decision made within the strategic framework. Try to foresee major potential problems, and address them on this basis.

Inclusivity does not mean that the strategic plan must be very detailed; it means that such details, when they become necessary, should fall within parameters or policies set in the strategic plan. The major part of the strategic plan is the definition and setting out of guidelines for the further, more detailed planning process. As project work is to the project plans, so the project plans are to the strategic plan.

There are many books available on the art of strategic planning, and no doubt the most widely read planners are at an advantage. However, a project is possibly one of the easier environments in which to carry out strategic planning, because by definition a project starts with defined objectives. Feasibility studies embrace the most basic target-setting aspects for the project. This creates a potential problem of its own: it is important not to be seduced by the preceding study when commencing a project. Take a second look. In conclusion, there are three major questions to ask for the purposes of strategic analysis.

29.1 Is the target correct?

This question breaks down into two: is the target attainable, and can we do better? We have discussed some aspects of unattainable targets in Chapter 7, The Contracting Environment, and will not repeat them here. If you proceed on a project knowing that the target will not be achieved (and there can be some very cynical reasons for doing this), you must also know that overall it will be less efficiently carried out than if the bad news is accepted up front and realistic planning is substituted.

The second part of the question always *has* to be asked, even when everything seems to be engraved in stone, otherwise the level of thinking is hardly strategic. Try brainstorming if necessary. It may be that the client has very set ideas about the target, and will not countenance any challenge. That is fine; just ensure that the potential opportunities or improvements, and the limitation on exploiting them, are recorded in the best way possible with least damage to the client relationship. Usually, any possibilities for improvement will surface again in the course of the project, and very damaging they can be at a later date.

If the answer to the second part of the question is positive, subsequent action is fairly obvious:

- evaluate the opportunities fully;
- verify that the improvements are real;
- decide how their introduction can best be utilized for 'our' benefit;
- negotiate the changes.

In the case of the project team undertaking work in accordance with client specifications, obviously the probing and full understanding of the specifications/contract is an essential prerequisite to this part of the strategic planning process.

29.2 How do we get there?

Engineering work is guided by the design criteria, which (after defining the design objectives) outline how the design will be performed, rather than the design itself. At the strategic level we need to ensure that the criteria are appropriate to every important aspect of our overall project objectives.

- Decide whether site visits, visits to other plants, or expert assistance is needed before finalizing the design criteria or proceeding with any work.
- Choose the design criteria checklist (for example Appendix 2).
- Define the criteria that will influence the answers to the checklist questions, for instance the following.
  - Engineering is to be carried out at minimum cost, provided that a just-acceptable standard is maintained. (Appropriate for a contractor engaging in a lump-sum engineering job that will attract no further business.)
  - Plant costs are to be minimized, subject only to safety and ergonomic standards corresponding to health and safety regulations, regardless of any increases of operating cost. (For the investor who is tied to a very limited capital budget, but expects handsome operating profits.)
  - Plant capital and operating costs, and the cost of project engineering and management, are to be strictly optimized on the basis of 12 per cent per annum discount on future income and 5 per cent per annum inflation of product revenue and operating costs, for a plant life of 30 years.

At the project strategy level there are three further questions to be asked that heavily impact on engineering.

- What resources are needed? Do we have the right resources? Could we do better?
- What is the best way of 'packaging' the project work at a high level, for example:
  - sub-contract large design-and-construct turnkey packages, thereby cutting down on engineering and management costs;
  - do everything piece-small and maximize the project team's own input.

  This decision will be affected strongly by the existence or otherwise of any restrictions or objectives as to the ratio of IFC to DFC.
- What is the procurement strategy (including the procurement of construction work)? See Chapters 6 and 19.

29.3 What are the major problems?

This question, and the action plan arising, should have been addressed during risk analysis in the feasibility study (Chapters 9 and 12), but may need to be repeated in more depth at the project stage. The high-level risk management plan is part of the strategic plan.

In conclusion: already too much has probably been said about strategic planning – the whole idea is to think big, and this is not necessarily facilitated by an excess of advice.

Chapter 30

Key Issues Summary

This is a summary emphasizing some of the most important observations and conclusions from the preceding chapters. The issues listed here are important generalities of approach to the job; it should not be inferred that all potentially critical issues are covered. Rather, the list should be customized at the outset of each project. It is intended as an *aide-mémoire* for the engineer to take stock at various intervals as the project proceeds.

30.1 Key project or study issues at the conceptual stage

- Developing an understanding of the client culture, and in particular his expectations in regard to design standards and innovation.
- The application of lateral thinking and value engineering. Use as much ingenuity as can safely be exercised, particularly by the use of proven techniques in new applications.
- The use of established and defensible estimating techniques. Specifically identify the basis for cost allowances and contingency. Avoid making firm commitments except on a basis of adequate engineering, firm prices from the market-place, and thorough risk analysis. Be critical in the use of data based on previous projects, especially those set up as self-fulfilling prophesies.

30.2 Key issues at the project execution phase

- Planning and working to plan. The most efficient performance is obtained by dealing with all conceptual issues up front, planning the work in successive levels of detail, and sticking to the plans except for minimized and formally controlled change. Adequate planning includes wide-ranging strategic review, identifying critical items, anticipating problems, and the issue of work procedures whose mandatory content includes only what is really necessary. Clear definition and verification of project scope is an essential part of establishing the baseline of the plan.

- Maintaining the right balance between time, cost and technical quality. Do not allow short-sighted individual aspirations (involving any one aspect) to predominate.
- The making and control of assumptions. In a competitive environment, making a certain amount of assumptions is essential to break the chains of unaffordable logical sequence. Exercise control by following through to verify the assumptions and by understanding and allowing for the consequences of wrong assumptions.
- Balancing of work input and work quality. There is good engineering and bad engineering, but no such thing as engineering perfection. A minimal amount of errors entails many iterations, which become unaffordable. Identify the critical issues that have to be right, and for the rest maintain a balance between extra engineering cost and the consequences of error. Let neither perfectionists nor false economists predominate.
- Controlling consultants and any other external experts (including the clients') who participate in the design process. To do this properly requires a strategy for each relationship by which there is an advance commitment (or 'buy-in') to achieve mutually acceptable objectives.
- Maintaining a good commercial interface. This requires a comprehensive strategy linking the goals of design and procurement, and a system of purchasing specifications that is value- and market-orientated.
- Development of sound layouts, which are exhaustively reviewed with the use of appropriate checklists.
- Control of plant-space. Keep design development under constant review, and note how each item of space is occupied by each discipline. Identify operator and maintenance space up front, and prevent encroachment. If you are not in control of the plant-space, you are not in control of the project.
- Staying in control of the project. This requires rigid document control, prompt dealing with problems as they arise, and maintaining the initiative in contractual relationships (which

involves forethought and planning). If there are signs of impending chaos (loss of control), take drastic action immediately. This may ultimately entail renegotiating the workscope or the time-frame.
- Plan backwards as well as forwards. Have a defined close-out plan from the outset, including defined close-out documentation, and plan backwards from this point as well as forwards from the present.

30.3 Key issues for the conduct of the individual engineer or engineering business unit

At the study or conceptual stage

- Target-setting and acceptance. Do not be manipulated into accepting targets that are not both adequately defined and attainable (and do not include weasel-words which may hide extra work).
- Contingency management. Do not allow certain cost items of presently uncertain value to be sucked into a general contingency. Insist that such items are included in specific 'line-item' allowances. Ensure that the actual contingency includes an allowance, based on experience or statistical analysis, for genuinely unforeseen items, and fight for as much discretionary contingency as possible while minimizing that of sub-contractors.
- Do not accept liabilities which are out of proportion to the work performed, or inconsistent with the proposed project scenario (for example based on engineering perfection in a limited-cost and limited-time environment).

At the project execution stage

- If unforeseen developments make it impossible to continue the project without maintaining quality control that is appropriate to the responsibilities accepted, halt the issue of all potentially suspect documentation until the situation is rectified by firm management action. Do not be sucked into a situation of continuing chaos.
- Keeping the initiative. Engineering, by its creative nature, creates opportunities for seizing the initiative; take it and hold on to it.

30.4 General business issues

- Management of important relationships (with clients, sub-contractors, partners, and so on). Understand what behaviour is acceptable in the cultural climate of the project. Be observant: large project expenditures attract rogues with a predatory agenda concealed beneath an obliging exterior. Spot the rogues and develop a strategy to deal with them.

- Nurturing of quality clients. Recognize the quality clients and strive to give them unbeatable service. Recognize unreasonable clients, and develop a plan to limit your services to the minimum of your obligations. And if your clients are all unreasonable, what are you doing in this business anyway?

Appendix 1

Jargon

As in most industries, many of the key words used have meanings or connotations not included in the standard dictionary. Sometimes the special connotations are subtle; sometimes a whole world of industrial practice is implied. An understanding of the special usage and interpretation of these words is essential to communication in the industry.

Authors of professional and technical literature, especially those striving to promote a new discipline, frequently propose a new 'jargon' to name original concepts. The jargon is considered to be essential to label some thing or activity to make it a subject for everyday usage and recognition within that field. In the case of project management and engineering, the jargon includes relatively few specially coined words, but rather many standard words which in the context of the industry have acquired special meaning and usage. In general the terminology has evolved rather than been coined by an acknowledged author. Although there are various institutions and bodies which aspire to stamp some form of authority on this field, and with it standardized terminology, there has been no general acceptance. Many words have a usage that is generally understood, but there is also some confusion.

Therefore it is often necessary to define words within a project context, practically as part of the project procedures, especially when multicultural interfaces are involved. Words which require particular care include those used to define the performance parameters of plant and equipment, such as 'normal', 'design', and 'maximum' (we will return to this subject), and those which are involved in the relationship with a supplier or sub-contractor, such as 'approval'. Procurementrelated vocabulary definitions are a standard part of general conditions of purchase and contract: the project engineer should review such definitions and their adequacy before proceeding with procurementrelated activities, such as the preparation of specifications. Also requiring particular care are the words 'contingency', 'accuracy', and in fact many of the words involved in defining the breakdown of a cost estimate (see Chapter 9).

There are some positively horrible verbal concoctions in widespread use for process plant work, for instance 'deliverables', which is intended to mean those products of engineering and management work which can be delivered as documentation. No apology is offered for the omission of such slaughter of the English language from these pages, but this is a widely used term that needs to be understood.

A1.1 Definition of design parameters

There are some issues of capacity definition which are often the cause of confusion, with regard to individual plant items as well as to the plant as a whole. Note that individual plant items (equipment, pipes, conveyors, etc.) usually need the facility to function at capacities higher or lower than the theoretical normal in order to allow for surge conditions, process variations, uncertainties, and so on. There are many national, industry, and equipment standards (in the context of their application) that define the usage of adjectives such as 'normal', 'maximum', 'rated', and 'design', when used to qualify performance parameters such as capacity, pressure, and power absorption. There is no overall standard for the process plant industry. The most frequent confusion arises with respect to the usage of 'design', which can be confused as being either 'normal' or 'maximum' (especially by an equipment vendor explaining why his equipment has failed!).

The project engineer must ensure that these terms are used in the context of a standard which defines them, or that a clear definition is given, and that there is a common understanding among the project participants. As examples of 'project definitions', consider the following.

- 'Maximum rated' conditions are the maxima for which continuous plant operation is required. All plant equipment shall be guaranteed for continuous operation at maximum rated capacity while subjected to the most severe possible combination of ambient and other conditions.
- 'Normal' conditions are the conditions anticipated for most frequent operation, and for which optimum plant efficiency is needed.
- 'Design' is not to be used to define specific plant operating conditions, except in the context of a referenced standard or code which defines the word.

Appendix 2

Design Criteria Checklist

A well-drafted set of design criteria will provide a balanced and comprehensive guide to the development of engineering work, in a way which exactly meets the performance requirements of the plant and the objectives of the project. The design criteria should address the plant as a whole, such that the design stages which follow augment the detail without the need to revisit the fundamental concepts. Thus the design criteria should commence with the definition of plant performance (including all the aspects of performance, such as operability and maintainability, which may determine 'fitness for purpose'); should reference any special, local, or environmental features which may need to be considered for plant design; and should determine the design methods and standards to be employed. The following are some notes on the essential content.

A2.1 Plant performance

- The process plant product specification. (What comes out of the plant.) This must clearly include not just target values of parameters, but also the limiting values by which the plant product will be accepted or rejected. It is important to be pedantic and try to consider all parameters and characteristics which may possibly be relevant: all too many projects have failed because certain parameters (such as moisture content, exclusion of unacceptable substances, or size grading) were either overlooked or taken for granted.
- The specification of the feedstock and imported utilities such as water, power, and chemical reagents. (What goes into the plant.)

The same care is required as in the case of the product. In the case of purchased utilities or reagents there may be leeway for the project engineers to determine the specifications, but there will inevitably be limitations (for example of price or source) which should be ascertained at the outset.

- Limitations on plant effluent, waste disposal, and environmental impact.
- The plant capacity (its rate of working). This may be defined in terms of quantity of product, quantities of types of product, quantity of feedstock, or a combination of quantities to be produced and processed over a period of time.

  'Period of time' needs to be qualified, to take into account: the plant operational and maintenance requirements; whether operation is continuous or how many hours per week or year; and the time allowed for scheduled and unscheduled shutdowns for maintenance, inspection, and repair (collectively referred to as 'maintenance'). The starting point for plant design is usually the 'average' capacity (over a long period of time which includes all shutdowns for whatever reason). A 'normal' capacity is then calculated for the time when the plant is in operation, allowing for shutdowns according to the defined plant operational mode and the time to be allowed for maintenance; these have a major bearing on the plant design and reliability requirement.
- The required flexibility of operation. The range of capacities over which the plant must be operable (or the 'turndown') is the most important consideration, but all required abnormal operation should be considered, including initial start-up, normal start-up, normal shutdown, emergency shutdown, and power failure. Sometimes a 'maximum' capacity may be defined to provide an extra margin for operational contingencies; however, care must be exercised in the use of this term, as previously outlined.
- Plant reliability, maintainability, and life requirements, consistent with the plant capacity calculation.
- Requirements for the operation of the plant, limitations on the numbers and skills of operational personnel, degree of automation, and local or statutory regulations which may affect plant operation.
- Special safety requirements, hazard containment, and fire prevention and extinguishing.
- Plant performance testing and acceptance standards.

A2.2 Plant site

- The site location and document references for topographical and survey information, describing the land available for the plant and for construction facilities, adjoining plant or structures, and access.
- Site ambient conditions (including altitude, maxima and minima of atmospheric conditions, atmospheric corrosion, wind, and dust).
- 'Battery limit' definition and information. Location of points where feedstocks, utilities, reagents, etc. are received, where products are despatched, and up to which roads and site facilities extend.
- Soils report for foundation design.
- Hydrological data for stormwater run-off design.
- Seismic design requirements.
- Local and environmental restrictions that may affect plant construction.

A2.3 The process

This section may be a document of its own, for example a process technology manual or a set of instructions from a licensor. Refer to Chapter 2 for a general description of process package contents. Whereas the Plant Performance section addressed what must be achieved, this section addresses how the process plant will be configured to achieve it. Apart from the choices and methodologies arising directly from the previous section, the subjects to be addressed include:

- calculation basis for mass and heat balance;
- capacity and other factors to be employed to determine the required performance of individual plant equipment items;
- criteria and methods to be employed for sizing static plant items such as vessels, stockpiles, and pipelines;
- identification of hazards, corrosive substances, and other characteristics within the process for which special care is required when designing the plant (plus details of established practice for dealing with these hazards);
- process design codes to be employed;
- design basis for plant instrumentation and control (often referred to as the 'control philosophy').

The basis for process calculations, special requirements arising out of process needs, and process design decisions should be listed and updated for the duration of the project.

A2.4 Basic plant design features

- Allowance of space, plant configuration, connections, over-capacity of equipment, or other design features to allow for future plant expansion or possible modification.
- Statutory and government authority regulations and approvals, insurers' and process licensors' requirements which may affect plant design.
- Corrosion protection systems.
- Whether plant units will be enclosed from the atmosphere and, if so, type of enclosure.
- Personnel access and ergonomic standards for operation and maintenance. Systems and features for removal and replacement of equipment.
- Size and space requirements for transport vehicles and mobile cranes: turning circles, clearance under structures, and so on.
- Plant drainage system and implications for unit elevations.
- Prevailing winds and implications for plant layout.
- Design practice for installed standby equipment.
- Design practice for emergency power requirements.
- Facilities for plant operators and maintenance personnel.

A2.5 Design features for equipment (mechanical, instrumentation, and electrical)

- Applicable international, national, and other standards.
- System to ensure maximum economic commonality of spare parts.
- Service factors to be employed, such as for sizing of electric motors and power transmissions.
- Features required for plant lubrication, maintenance, and replacement, including vendor support.
- Any other features which previous plant experience has shown to be desirable.

A2.6 Design methodology and standardization

- Standard design codes, methods, software, designs, bulk material components, and construction materials to be employed for all disciplines, for example structures, piping, and buildings.

- System of measure, such as SI units.
- Methodology for hazard and operational reviews, and design for plant safety.
- Drafting and modelling standards and software.
- Methodology for materials take-off, presentation of bills of quantities, and materials control, in so far as it affects the design documentation.
- Numbering systems for documents, equipment, and code of accounts.

A2.7 Design verification and approval plan

- Within project team.
- By process technology supplier (if external to the team).
- By client.
- By external regulators.
- Needs for each engineering discipline.

A2.8 Design and plant documentation for client

- Language(s) to be used on official documents.
- Client requirements for format and presentation of documents and computer disks.
- Content and presentation of plant operating and maintenance manuals.
- Project close-out documentation.

Index

Printed and bound by CPI Group (UK) Ltd, Croydon, CR0 4YY

16/04/2025

14658823-0001